Lecture Notes
in Control and Information Sciences 221

Editor: M. Thoma

Springer-Verlag London Ltd.

Atul Kelkar and Suresh Joshi

Control of Nonlinear Multibody Flexible Space Structures

Springer

Authors

Dr Atul G. Kelkar
Department of Mechanical Engineering
Kansas State University, Manhattan, Kansas 66506, USA

Dr Suresh M. Joshi
NASA Langley Research Center, Hampton, Virginia 23681, USA

ISBN 978-3-540-76093-1 ISBN 978-3-540-40944-1 (eBook)
DOI 10.1007/978-3-540-40944-1

British Library Cataloguing in Publication Data
Kelkar, Atul G.
Control of nonlinear multibody flexible space structures -
(Lecture notes in control and information sciences ; 221)
1.Large space structures (Astronautics)
I.Title II.Joshi, Suresh M.
629.4'7

Library of Congress Cataloging-in-Publication Data
a catalog record for this book is available from the Library of Congress

Typesetting: Camera ready by author
69/3830-543210 Printed on acid-free paper

To my mother and the memory of my father
– A.G.K.

To my mother and father
– S.M.J.

"The reasonable man adapts himself to the world; the unreasonable one persists to adapt the world to himself. Therefore all progress depends on the unreasonable man."

George Bernard Shaw

Preface

A number of space missions such as astronomy, Earth observation, and communications require spacecraft with multiple articulated appendages. The performance requirements often dictate high-precision control of the spacecraft attitude as well as the appendage motion. The dynamics of such systems are inherently nonlinear, and the presence of significant elasticity in the members and the joints further complicates the problem of control system design for such systems. An important subclass of multibody flexible space systems consists of autonomous space-based robots. Future exploration and utilization of space, as well as interplanetary travel, will require such systems for automated construction and assembly in space and on other planets. These tasks are not only hazardous to humans, but would also require high precision beyond human capability. The availability of space robots would drastically reduce the necessity of extra-vehicular human activity. Because of the presence of flexibility, modeling inaccuracies, and inherent nonlinearities in the dynamics, control system design for such systems would be a difficult challenge, and is the motivation for this book.

This book is a follow-on of the book: "Control of large flexible space structures" by the second author, which was published in 1989 as volume 131 in this series. The first book mainly addressed the problems of fine attitude pointing and vibration suppression for a class of flexible spacecraft that do not have articulated appendages, i.e., "single-body" spacecraft. Because fine-pointing and vibration suppression problems are essentially "small motion" or linear problems, a wide variety of control design methods are available to address them. The important requirement, however, is that the control system must be ***robust*** in the presence of model uncertainties, nonlinearities, and failures, and these issues were addressed in the previous book for linear, single-body flexible space structures.

The present book addresses the logical next step, i.e., the problem of robust control of multibody flexible space systems. This is an important practical problem for the reasons mentioned previously. The main result presented is a class of output feedback controllers that offer globally stable closed-loop maneuvers for nonlinear multibody flexible spacecraft. The control laws are robust to parametric uncertainties, modeling inaccuracies, unmodeled dynamics, and in some cases, actuator and sensor nonlinearities. The basic tools used for obtaining the stability proofs are passivity theory and Lyapunov-LaSalle methods.

The control laws are also applicable to terrestrial robot control as well as to globally stable closed-loop large-angle maneuvering of single-body spacecraft.

Several researchers have worked on this important problem during the past three decades. Although we have attempted to give an adequate bibliography, it is very possible that it may still be incomplete, and we would like to apologize for any inadvertent omissions. The results presented in this book constitute only one step towards the solution of this important problem, and much work, both analytical and experimental, needs to be done in order to solve this problem. It is our hope that this book will help stimulate research activity in this problem and will ultimately lead to its complete solution and practical implementation.

Atul G. Kelkar
Manhattan, Kansas
April 1996

Suresh M. Joshi
Hampton, Virginia

Contents

Chapter 1

Introduction

Space missions such as multi-payload space platforms and space-based manipulators will utilize flexible space structures (FSS) in low-Earth as well as geosynchronous orbits. Examples of near-term missions involving flexible structures include Earth observing systems and manipulators for automated on-orbit assembly and satellite servicing. Future space mission concepts include mobile satellite communication systems, solar power satellites, and large optical reflectors, which would require components such as large antennas, platforms and solar arrays. Such systems would typically range in dimension from 50 meters to possibly several kilometers. Their relatively light weight and in some cases, expansive sizes, would result in several low-frequency, lightly damped structural (elastic) modes. The natural frequencies of the elastic modes would be generally closely spaced, and some frequencies may lie within the controller bandwidth. Furthermore, the elastic mode parameters would not be known accurately. These characteristics make the control systems design for flexible space structures a difficult problem.

Flexible spacecraft can be roughly categorized according to their missions as single-body spacecraft and multibody spacecraft. Two important control problems for single-body FSS are: i) fine-pointing in space with the required precision in attitude (i.e., orientation) and shape, and ii) "slewing" or large-angle maneuvering to orient to a different target. Both of these problems usually have very high performance requirements. As an example of performance specifications, a certain mobile communication system concept consisting of a 120-meter diameter space-antenna will have pointing accuracy requirement of 0.03 degree root mean square (RMS). Requirements for certain other missions are expected to be even more stringent, on the order of 0.01 arc-second. Some missions would have the requirement to quickly maneuver the FSS through large angles in order to rapidly acquire a new target on Earth, while ensuring minimum fuel expenditure. The elastic motion and the accompanying stresses must be kept within acceptable limits during the maneuver. After acquiring the target, the FSS must point to it with specified precision.

The main control problems for multibody spacecraft are: i) fine-pointing of some of the appendages to different targets, ii) rotating some of the appendages to track specified periodic scanning profiles, and iii) changing the orientation of some of the appendages through large angles. For instance, multi-payload platforms would have the first two requirements, whereas multi-link manipulators would have the third requirement in order to reach a new end-effector position.

The distinguishing feature that separates FSS from conventional older-generation spacecraft is their special dynamic characteristics caused by their highly prominent structural flexibility. Literature surveys on dynamics and control of FSS may be found in [Nur.84] and [Hyl.93].

This book is intended to be a sequel to an earlier book [Jos.89] by the second author, which addressed in detail the problem of fine-pointing control of single-body FSS. The single-body FSS problem not only represents an important class of missions, but also permits analysis in the linear, time-invariant (LTI) setting. We shall next present a brief summary of the results presented in [Jos.89] for single-body FSS.

1.1 Linear Mathematical Models of Single-Body FSS

Flexible space structures are basically infinite-dimensional systems. It is possible to model simple flexible structures, such as uniform beams or plates, by infinite-dimensional systems (see [Jos.89]). Approximate infinite dimensional models also have been proposed for more complex structures such as trusses [Bal.92]. Most of the realistic FSS, however, have a highly complex geometry and not amenable to infinite-dimensional modeling. The standard engineering practice is to use finite-dimensional mathematical models generated by using the finite element method [Mei.70]. The basic approach of this method is to divide a continuous system into a number of elements using fictitious dividing lines, and to apply the Lagrangian formulation to determine the forces at the points of intersection as functions of the applied forces. Suppose there are r force actuators and p torque actuators distributed throughout the structure. The ith force actuator produces the 3×1 force vector $f_i = (f_{xi}, f_{yi}, f_{zi})^T$, along the X, Y, Z axes of a body-fixed coordinate system centered at the nominal center of mass (c.m.). Similarly, the ith torque actuator produces the torque vector $T_i = (T_{xi}, T_{yi}, T_{zi})^T$. Then the linearized equations of motion can be written as [Jos.89]:

Rigid-body translation:

$$M\ddot{z} = \sum_{i=1}^{r} f_i \tag{1.1}$$

Rigid-body rotation:

$$J\ddot{\alpha} = \sum_{i=1}^{r} R_i \times f_i + \sum_{i=1}^{p} T_i \tag{1.2}$$

Elastic motion:

$$\ddot{q} + D\dot{q} + \Lambda q = \sum_{i=1}^{r} \Delta_i^T f_i + \sum_{i=1}^{p} \Phi_i^T T_i \tag{1.3}$$

where M is the mass, z is the 3×1 position of the c.m., R_i is the location of f_i on the FSS, J is the 3×3 moment-of-inertia matrix, α is the attitude vector consisting of the three Euler rotation angles (ϕ, θ, ψ), and $q = (q_1, q_2, \ldots, q_{nq})^T$ is the $n_q \times 1$ modal amplitude vector for the n_q elastic modes. ("$\times$" in (2) denotes the vector cross-product). The number of modes (n_q) necessary for adequately characterizing an FSS is usually quite large, perhaps 100-1000. Δ_i^T, Φ_i^T denote the $n_q \times 3$ translational and rotational mode shape matrices at the ith actuator location. The rows of Δ_i^T, Φ_i^T represent the X, Y, Z components of the translational and rotational mode shapes at the location of the ith actuator, respectively.

$$\Lambda = diag(\omega_1^2, \omega_2^2, \ldots, \omega_{n_q}^2) \tag{1.4}$$

where ω_i represents the natural frequency of the ith elastic mode and D is an $n_q \times n_q$ matrix representing the inherent damping in the elastic modes:

$$D = 2\,\mathrm{diag}\,(\rho_1\omega_1, \rho_2\omega_2, \ldots, \rho_{nq}\omega_{nq}). \tag{1.5}$$

The inherent structural damping ratios (ρ_i's) are typically on the order of 0.001-0.01. The inherent damping cannot be modeled by the finite element method, and it is customary to add a proportional damping term after an undamped finite element model is obtained.

The translational positions and rotation angles at a location with coordinate vector R, are given by:

$$z_p = z - R \times \alpha + \overline{\Delta} q \tag{1.6}$$

$$y_p = \alpha + \overline{\Phi} q. \tag{1.7}$$

(The ith columns of $\overline{\Delta}$ and $\overline{\Phi}$ represent the ith 3×1 translational and rotational mode shapes at that location.)

The mode-shape plots for the 122 m diameter, hoop/column antenna concept [Rus.80], which consists of a deployable central mast attached to a deployable hoop by cables held in tension, are shown in Figure 1.1.

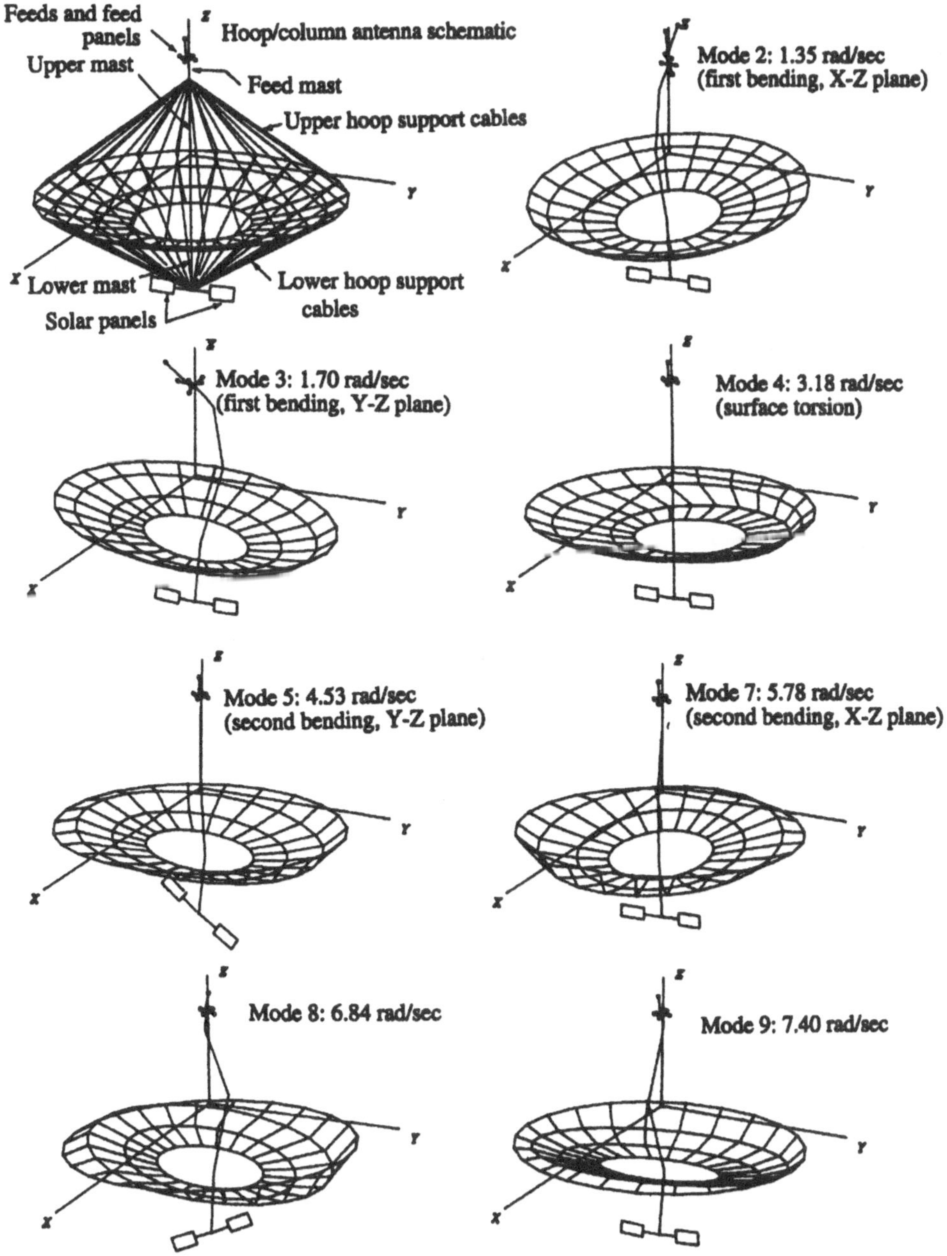

Figure 1.1: Typical mode shapes for hoop/column antenna

1.2 Controllability and Observability

Fine attitude pointing control is usually accomplished by using torque actuators rather than force actuators [Jos.89]. Therefore, only rotational equations of motion need to be considered to study this problem, and no force actuators are used. It was shown in [Jos.89] for the single-body fine-pointing problem that the system is controllable if and only if (iff)

i) the row of the mode-shape matrix Φ^T corresponding to each distinct (in frequency) elastic mode has at least one nonzero entry, and

ii) if there are ν elastic modes with the same natural frequency $\overline{\omega}$, the corresponding rows of the mode-shape matrix form a linearly independent set.

The first condition is satisfied iff the rotational mode shape for each mode is nonzero at the location of at least one actuator. The second condition is needed only when there are more than one elastic modes with the same natural frequency, a common occurrence in symmetric structures. Similar necessary and sufficient conditions are available for observability. It is important to note that the rigid-body modes are not observable using attitude-rate sensors alone without attitude sensors. However, a three-axis attitude sensor (without rate sensors) can be sufficient for observability [Jos.89].

1.3 Controller Considerations

Precise fine-pointing control requires controlling the rigid rotational modes and suppressing the elastic vibration. The control objectives for fine pointing are:

1) **Rapid transient response-** Quick damping out of the pointing errors caused by abrupt disturbances (such as thermal distortion due to entering or leaving Earth's shadow), or nonzero initial conditions (e.g., resulting from the completion of a large-angle attitude maneuver), and

2) **Disturbance rejection-** Maintaining the attitude near the desired attitude in the presence of noise and disturbances.

To accomplish the first objective, the closed-loop bandwidth must be greater than a specified value. The second objective translates into minimizing the RMS pointing error. In addition, the permissible elastic motion is usually very small; i.e., the RMS shape distortions must be below prescribed limits. Some applications (e.g., large communications antennas) typically require closed-loop bandwidth of at least 0.1 rad/sec, with a maximum of 4 second (closed-loop) time constant for all the elastic modes. Typical permissible RMS errors are: 0.03 deg. pointing error, and 6 mm surface distortion.

The following problems are encountered in designing a precision attitude control system.

1) An acceptable model of an FSS is of high order because of large number of significant elastic modes; however, a practically implementable controller must have sufficiently low order.

2) The inherent damping is usually very small.

3) The elastic mode frequencies are low and often closely-spaced.

4) The model parameters, which include frequencies, damping ratios and mode shapes, are not known accurately.

A simple approach for controller design would be to truncate the model beyond a certain number of modes, and design a reduced-order controller. This approach is used for relatively rigid conventional spacecraft, wherein the design model consists only of the rigid modes. Low-pass filters are included in the loop to attenuate the contribution of the elastic modes. This approach is not appropriate for FSS because the elastic motion is much more prominent. Figure 1.2 shows the effect of using a truncated design model. When implementing a control loop around the "controlled" modes, an unintended feedback loop is also constructed around the truncated modes, which can make the closed-loop system unstable. This inadvertent excitation of the truncated modes by the input, and the undesired contribution of the truncated modes to the sensed output, are known as "control spillover" and "observation spillover" [Bal.82]. The spillovers may cause performance degradation, and can even destabilize the system.

Another problem in controller design is the lack of accurate knowledge

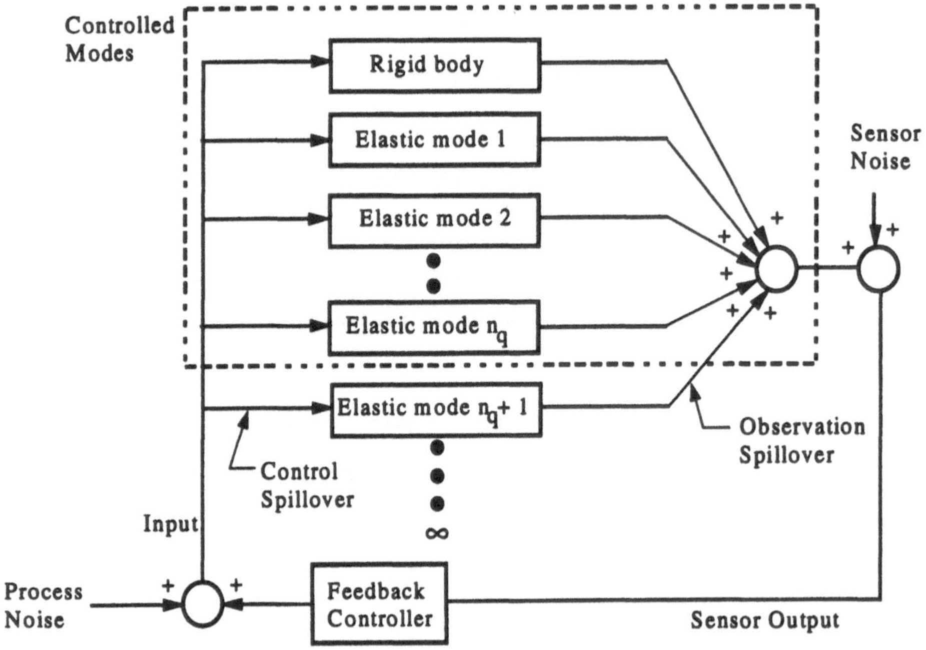

Figure 1.2: Control and observation spillover

of the parameters. Finite element models are known to give reasonably accurate estimates of the frequencies and mode shapes only for the first few modes, and can provide no estimates of inherent damping ratios. Ground-testing for parameter estimation is not generally feasible because many FSS cannot withstand the gravitational force while deployed, and some FSS may require very large test facilities such as vacuum chambers. Another controller design consideration is that the actuators and sensors have nonlinearities and finite bandwidth. Therefore, the attitude controller must be a "robust" one, that is, it must maintain at least stability, and perhaps performance, despite modeling errors, uncertainties, nonlinearities and component failures. Two important linear controller design approaches for the attitude control problem are model-based controllers, and model-independent passivity-based controllers.

1.4 Model-Based Controllers

The nth order state space model of an FSS, in the presence of process noise and measurement noise, can be expressed as

$$\dot{x} = Ax + Bu + v; \quad y = Cx + w \tag{1.8}$$

where $v(t)$ and $w(t)$ are respectively the $n \times 1$ and $l \times 1$ process noise and sensor noise vectors. v and w are assumed to be mutually uncorrelated, zero-mean, Gaussian white noise processes with covariance intensity matrices V and W. A linear-quadratic-Gaussian (LQG) controller can be designed to minimize

$$J = \lim_{t_f \to \infty} \frac{1}{t_f} \mathcal{E} \int_0^{t_f} [x^T(t)Qx(t) + u^T(t)Ru(t)]dt \tag{1.9}$$

where "$\mathcal{E}$" denotes the expectation operator, and $Q = Q^T \geq 0$, $R = R^T > 0$ are the state and control weighting matrices. The resulting nth order controller consists of a Kalman-Bucy filter (KBF) in tandem with a linear quadratic regulator (LQR), and has the form:

$$\dot{\hat{x}} = (A + BG - KC)\hat{x} + Ky; \quad u = G\hat{x} \tag{1.10}$$

where $\hat{x}$ is the controller state vector, $G_{m \times n}$ and $K_{l \times n}$ are the LQR and KBF gain matrices. Any controller using an observer and state estimate feedback has the same mathematical structure. The order of the controller is n, the same as that of the plant. An adequate model of an FSS typically consists of several hundred elastic modes. To be practically implementable, however, the controller must be of sufficiently low order. A reduced-order controller design can be obtained in two ways; either by using a reduced-order "design model" of the plant, or by obtaining a reduced-order approximation to a high-order controller. The former method is used more widely than the latter method because high order controller design relies on the knowledge of the high frequency mode parameters, which is usually inaccurate.

Several methods for model-order reduction have been developed during the past few years. The most important of these include the singular perturbation method, the balanced truncation method, and the optimal Hankel norm method (see [Gre.95] for a detailed description). In the singular perturbation method, higher-frequency modes are approximated by their quasi-static representation. The balanced truncation method uses a similarity transformation that makes the controllability and observability grammians equal and diagonal. A reduced-order model is then obtained by retaining the most controllable and observable state variables. The optimal Hankel norm approximation method aims to minimize the Hankel norm of the approximation error and can yield a smaller error than the balanced truncation method. A disadvantage of the balanced truncation and the Hankel norm methods is that the resulting (transformed) state variables are mutually coupled and do not correspond to individual modes, resulting in the loss of physical insight. A disadvantage of the singular perturbation and Hankel norm methods is that they can yield non-strictly proper reduced-order models. An alternate method of overcoming these difficulties is to is to rank the elastic modes according their contributions to the overall transfer function, in the sense of H_2, H_∞, or $\mathcal{L}_1$ -norms [Gup.94]. The highest ranked modes are then retained in the design model. This method retains the physical significance of the modes, and also yields a strictly proper model. Note that the rigid-body modes must always be included in the design model, no matter which order-reduction method is used. A model-based contorller can then be designed based on the reduced-order design model.

Reduced-Order LQG-Type Controller Design:- When a reduced-order design model is used to design an LQG controller, it stabilizes the design model. It may not, however, stabilize the full-order plant because of the control and observation spillovers. Some time-domain methods for designing spillover-tolerant reduced-order LQG controllers are discussed in [Jos.89]. These methods aim to reduce the norms of spillover terms $||B_tG||$ and $||KC_t||$, where B_t and C_t denote the input and observation matrices corresponding to the truncated modes. Sufficient conditions for stability based on Lyapunov's method are derived in terms of upper bounds on the spillover norms and are used as guidelines in spillover reduction. Controllers obtained using these methods are generally quite conservative, and also require the knowledge of the truncated mode parameters in

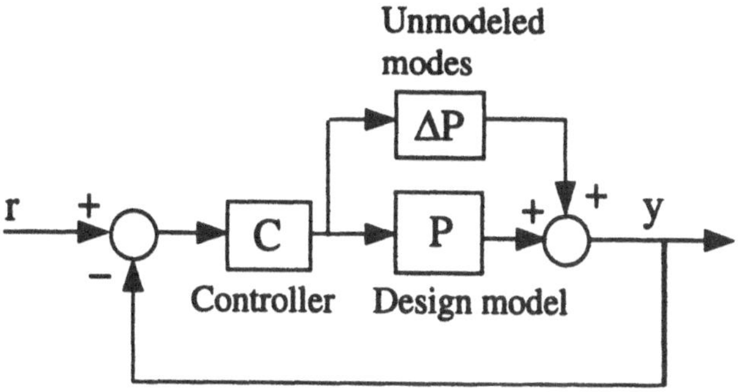

Figure 1.3: Additive uncertainty formulation of truncated dynamics

order to ensure stability.

The LQG controller design approach can be substantially improved by the application of multivariable frequency-domain methods. In this approach [Jos.89], the truncated modes are represented as an additive uncertainty term $\Delta P(s)$ that appears in parallel with the design model (i.e., nominal plant) transfer function $P(s)$, as shown in Figure 1.3. A sufficient condition for stability is [Gre.95]:

$$\overline{\sigma}[\Delta P(j\omega)] < \frac{1}{\overline{\sigma}[C(j\omega)\{(I + P(j\omega)C(j\omega)\}^{-1}]} \quad for \quad 0 \leq \omega < \infty \qquad (1.11)$$

where $\overline{\sigma}[.]$ denotes the largest singular value and $C(s)$ denotes the controller transfer function. An upper bound on $\overline{\sigma}[\Delta P(j\omega)]$ can be obtained from approximate knowledge of the truncated mode parameters, to generate an "uncertainty envelope". The stability test (1.11) is shown in Figure 1.4 for a typical large space antenna where the design model consists of the three rigid rotational modes and the first three elastic modes. The bandwidth of the closed loop transfer function $G_{CL} = PC(I + PC)^{-1}$ (Figure 1.5) represents a measure of the nominal closed-loop performance. The controller used in Figures 1.4 and 1.5 was obtained by iteratively designing the KBF and the LQR to yield the desired closed-loop bandwidth while still satisfying (1.11). The iterative method, which is described in [Jos.89], is loosely based on the LQG/Loop

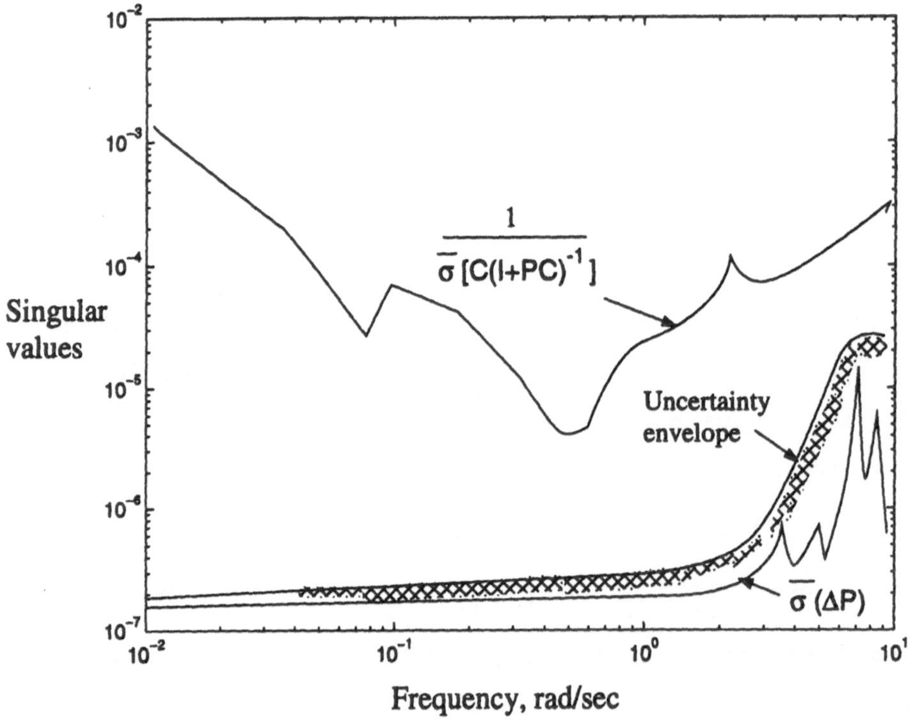

Figure 1.4: Stability test for additive uncertainty

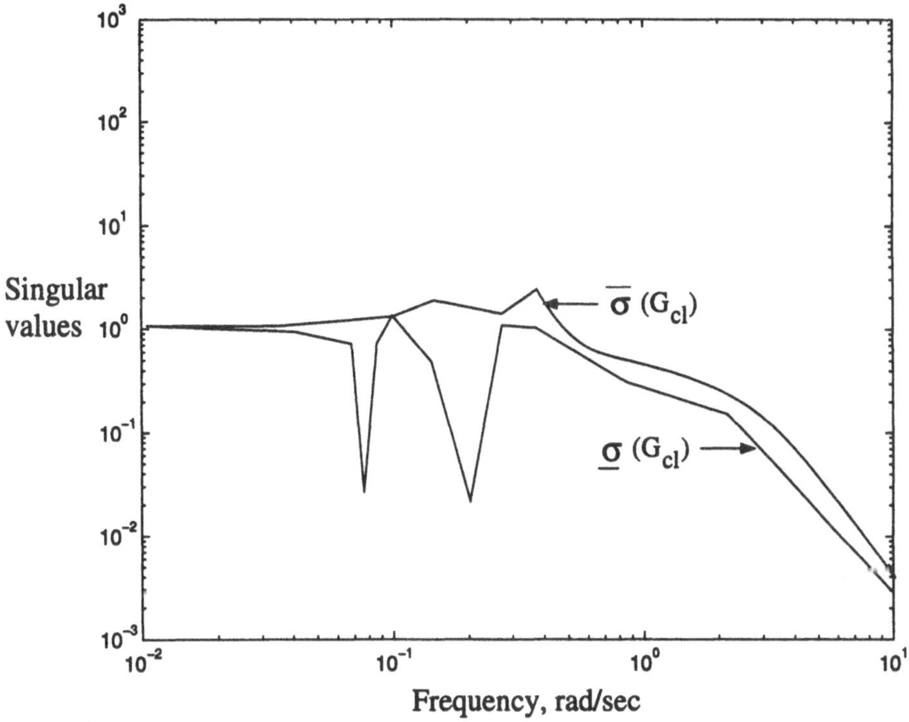

Figure 1.5: Closed-loop transfer function

Transfer Recovery (LTR) method [Ste.87]. The resulting controller is robust to any truncated mode dynamics which lies under the uncertainty envelope. The controller may not, however, provide robustness to parametric uncertainties in the design model. In particular, small uncertainties in the natural frequencies of the design model (i.e., the "controlled modes") can cause the condition (1.11) to be violated because of the sharp peaks in the $\overline{\sigma}[\Delta P]$ due to small open-loop damping. That is, small errors in the natural frequencies produce large error peaks in the frequency response [Jos.89].

H_∞ **and** μ**-synthesis Methods:-** A systematic method for obtaining the desired controller performance as well as robustness to truncated mode dynamics is the H_∞ method [Gre.95]. Typically, the design objective is to minimize the H_∞-norm of the frequency-weighted transfer function from the disturbance inputs (e.g., sensor and actuator noise) to the controlled variables, while ensuring stability in the presence of truncated modes. An example of the application of the H_∞ method to FSS control is given in [Lim.92]. The problem can be formulated to also include parametric uncertainties repesented as unstructured uncertainty; however, the resulting controller design is usually very conservative and provides inadequate performance.

The μ-synthesis or structured singular value method [Doy.82, Pac.93] can overcome the conservativeness of the H_∞ method. The procedure involves the "extraction" of the parametric uncertainties from the system block diagram and arranging them as a diagonal block that forms a feedback connection with the nominal closed-loop system. The controller design problem is formulated as one of H_∞-norm minimization subject to a constraint on the structured singular value of an appropriate transfer function. The problem can also be formulated to provide robust performance in the presence of model uncertainties. The application of the μ-synthesis method to FSS control is investigated in [Bal.94].

1.5 Model-Independent Controllers

If an attitude sensor and a rate sensor are collocated with each torque actuator, the input-output map from the torque vector to attitude-rate vector is passive, i.e., the corresponding transfer function matrix is positive-real [Jos.89]. This fact has been exploited to obtain a class of robustly stabilizing model-independent controllers, called ***static dissipative controllers***, given by the proportional-plus-derivative control law:

$$u = -G_p y_p - G_r y_r \tag{1.12}$$

where G_p and G_r are symmetric positive-definite matrices, and y_p, y_r denote the attitude and attitude rate vectors. The closed-loop system can be shown to be asymptotically stable regardless of the number of truncated modes or the knowledge of the parametric values; that is, the stability is ***robust***. The only requirements are that the actuators and sensors be collocated, and that the feedback gains be positive definite. Furthermore, if G_p, G_r are diagonal, then the robust stability holds even when the actuators and sensors have certain types of nonlinear gains. In particular, if

i) the actuator nonlinearities, $\psi_{ai}(\nu)$, are time-invariant monotonically non-decreasing and belong to the $(0, \infty)$ sector, i.e., $\psi_{ai}(0) = 0$, and $\nu\psi_{ai}(\nu) > 0$ for $\nu \neq 0$

ii) the attitude and rate sensor nonlinearities , belong to the $(0, \infty)$ sector, and the attitude-sensor nonlinearities are time-invariant,

then the closed-loop system with the static dissipative control law is globally asymptotically stable [Jos.89]. Examples of permissible nonlinearities are shown in Figure 1.6. It can be seen that actuator and sensor saturation are permissible nonlinearities which will not destroy the robust stability property.

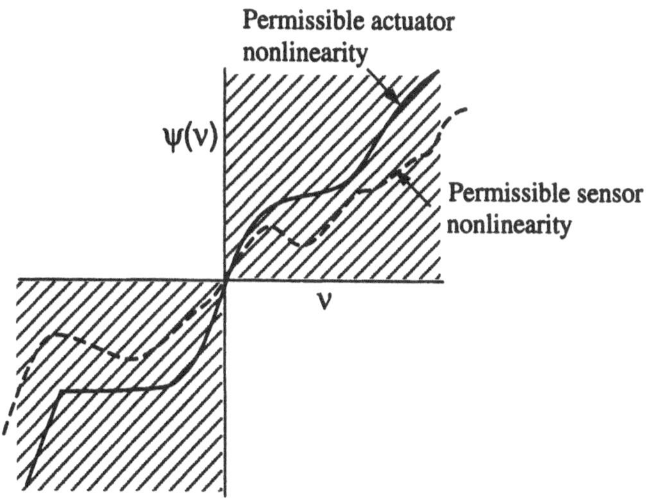

Figure 1.6: Permissible actuator and sensor nonlinearities

Some design methods for static dissipative contorllers are discussed in [Jos.89]. In particular, the static dissipative control law can be shown to minimize the quadratic performance function

$$\mathcal{J} = \int_0^\infty [y_p^T G_p G_r^{-1} G_p y_p + y_r^T G_r y_r + 2\dot{q}^T D \dot{q} + 2 y_p^T G_p G_r^{-1} u + u^T G_r^{-1} u]\, dt \tag{1.13}$$

which can be used as a basis for controller design. Another approach for selecting gains is to minimize the norms of the differences between the actual and desired values of the closed loop coefficient matrices.

1.6 Organization of This Book

The objective of this book is to present a class of stabilizing controllers for multibody flexible space structures which have inherently passive dynamics. The model-based controller design methods developed for linear FSS are not applicable to this problem. Therefore we employ a passivity-based approach. A mathematical model of a class of multibody flexible systems is derived in Chapter 2 using the Lagrangian formulation. The configuration is asumed to consist of a branched geometry, i.e., it has a central flexible body to which a number of articulated appendage bodies are attached. This configuration is very general and can adequately represent a multi-payload space platform or a multi-link robotic manipulator. The equations of motion are in the form of nonlinear coupled vector-matrix ordinary differential equations. With minor modifications (such as removal of the free-free modes and addition of the gravitational field), this model can also be used to represent terrestrial multilink flexible manipulators.

Chapter 3 presents the mathematical background required for stability investigation presented in the subsequent chapters. The concept of passivity is presented as a specialization of the more general concept of dissipativity with respect to a quadratic supply rate, and basic theorems for the stabilization of such systems are given. For linear, time-invariant (LTI) systems, passivity takes the form of positive realness. State-space characterizations of various classes of positive real systems are given, and sufficient conditions are obtained for the stabilization of linear and nonlinear passive systems by LTI positive-real compensators.

Chapter 4 addresses closed-loop control of nonlinear multibody flexible space structures. The central-body attitude is represented using the quaternion formulation, which allows unlimited rotational motion without kinematic singularities. Nonlinear control laws are presented, which employ the feedback of the central-body quaternions and angular velocity, as well as the joint angles and rates. These control laws are shown to provide global asymptotic stability. A special case, namely the "attitude hold configuration" occurs when the central-body motion is small but the appendage bodies can undergo unlimited motion. For this case, the system, which is still nonlinear, is shown to be stabilized by *linear* static dissipative controllers. It is shown that the static dissipative controller maintains stability in the presence of a broad class of actuator nonlinearities and actuator dynamics. A class of *dynamic* dissipative controllers is also introduced, which provides robust stability and potentially superior performance. The dynamic dissipative controller is also readily applicable to fine-pointing control of single-body FSS. In all cases, the stability is *robust* to unmodeled elastic modes and parametric uncertainties. The control laws presented permit *globally asymptotically stable* closed-loop maneuvers of multi-body or single-body FSS regardless of modeling errors. Numerical examples are presented for demonstrating the control of a nonlinear multibody FSS and a linear single-body FSS.

The problem of trajectory tracking for multibody FSS is addressed in Chapter 5. In particular, a nonlinear tracking control law is given for *rigid* multibody spacecraft, and is shown to provide asymptotically stable tracking of specified trajectories. This control law does not readily extend to multibody *flexible* spacecraft, and this remains a problem for future research. For the attitude-hold mode, however, a static dissipative tracking control law is shown to provide $\mathcal{L}_2$-stability. The results on tracking are preliminary, and much work needs to be done in order to obtain robust, high-performance tracking controllers.

Chapter 2

Mathematical Model

In this chapter, a mathematical model for a generic multibody flexible spacecraft is obtained. The spacecraft considered consists of a flexible central body to which a number of flexible multibody structures are attached. The coordinate systems used in the derivation allow effective decoupling of the translational motion of the entire spacecraft from its rotational motion about the center of mass. The derivation assumes that the deformations in the bodies are only due to small elastic motions. The dynamic model derived is a closed-form vector-matrix differential equation. The model developed can be used for analysis and simulation of many realistic spacecraft configurations.

2.1 Introduction

A class of next-generation spacecraft is expected to inlude nonlinear multibody flexible space systems. Some of the current spacecraft can also be catagorized under this class. Examples of such systems are [Ano.87, Asr.93, Jan.93]: satellites with flexible appendages, such as solar arrays and antennas, Space-Shuttle with remote manipulator system (RMS), and flexible space platforms with multiple articulated payloads. Mathematical modeling of these systems is quite complex. This problem has been addressed in the literature (e.g., see [Ng.88]); however, in this chapter, a different approach is taken to derive the equations of motion which yields a compact closed-form of the equations of motion. The derivation uses modeling techniques similar to those used in the robotics literature (e.g., see [Fu.87], [Spo.89]). The formulation is relatively general and can be used for a large class of spacecraft.

First, for the sake of completeness, some of the mathematical aspects of the modeling of rotating dynamical systems are summarized in the section on mathematical preliminaries. Next, the kinematic equations, i.e., the position, velocity, and acceleration relations, for a representative particle mass of the system are obtained. Once the kinematic equations are derived, the dynamics of the system can be modeled by using various methods; for example, the Newtonian approach, the calculus of variations approach, and the Lagrangian approach. The equations of motion derived using any of these methods are equivalent; however, the Lagrangian formulation is used here since it is an energy-based approach (i.e, it uses a scalar formulation) and is easy to work with compared to the Newtonian approach which deals with vector quantities. It is assumed in the derivation that the bodies deform only due to the elastic motion and the deformations are in the linear range. However, any other deformations such as deformations due to the thermal effects can be easily included in the formulation with some modifications in the potential energy function.

In deriving the kinematic equations for the chain of multiple flexible bodies, the coordinate systems become an important element of the derivation. A large part of kinematics deals with the coordinate transformations used to represent the position and orientation of the body. In view of this, we will begin with the operations of translation and rotation, and the transformations which are used to represent these motions.

2.2 Mathematical Preliminaries

2.2.1 Rotations

Consider a rigid body as shown in Figure 2.1, to which a body-axis system, i.e., a body-fixed coordinate frame, (X_b, Y_b, Z_b), is attached. Let (X_o, Y_o, Z_o)

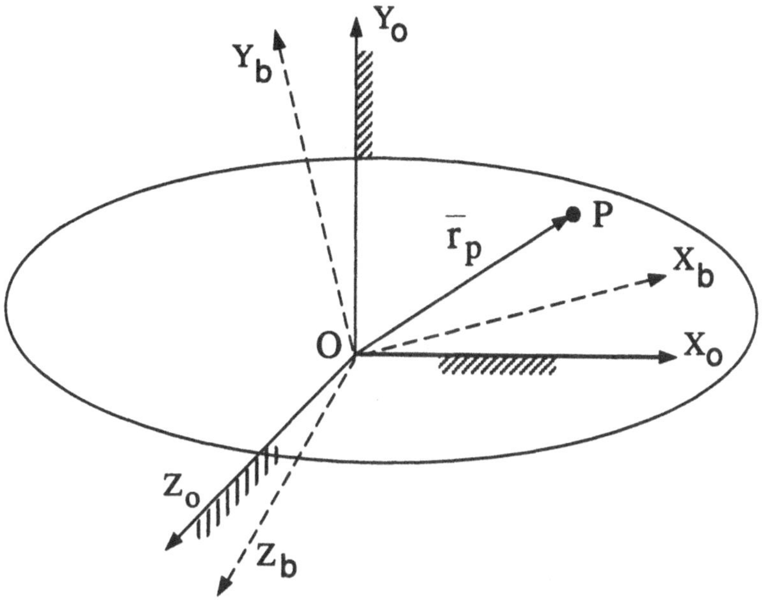

Figure 2.1: Fixed reference and body axes system

represent some fixed reference frame whose origin is concentric with the body axes system. Our aim is to relate the coordinates of a point P on the body in the (X_b, Y_b, Z_b) frame to those in the (X_o, Y_o, Z_o) frame. Let i_b, j_b, k_b denote orthonormal basis vectors in the body frame and i_o, j_o, k_o denote the orthonormal basis vectors in the fixed frame. Then, the position vector of point P, $\overline{r}_p$, in the body frame is given as

$$\overline{r}_p^b = r_{bx} i_b + r_{by} j_b + r_{bz} k_b \tag{2.1}$$

and in the fixed frame is given by

$$\overline{r}_p^o = r_{ox} i_o + r_{oy} j_o + r_{oz} k_o. \tag{2.2}$$

Since $\overline{r}_p^b$ and $\overline{r}_p^o$ are representations of the same vector $\overline{r}_p$, the relation between the components of $\overline{r}_p$ in two different coordinate systems can be obtained as follows.

$$r_{bx} = i_b \cdot \overline{r}_p^b = i_b \cdot \overline{r}_p^o. \tag{2.3}$$

Similarly,

$$r_{by} = j_b \cdot \overline{r}_p^b = j_b \cdot \overline{r}_p^o \tag{2.4}$$

and

$$r_{bz} = k_b \cdot \overline{r}_p^b = k_b \cdot \overline{r}_p^o. \tag{2.5}$$

Equations (2.3)-(2.5) can be rewritten in a compact form as

$$\begin{bmatrix} r_{bx} \\ r_{by} \\ r_{bz} \end{bmatrix} = R_o^b \begin{bmatrix} r_{ox} \\ r_{oy} \\ r_{oz} \end{bmatrix}$$

or, notationally,

$$\overline{r}_p^b = R_o^b \overline{r}_p^o \tag{2.6}$$

where

$$R_o^b = \begin{bmatrix} i_b \cdot i_o & i_b \cdot j_o & i_b \cdot k_o \\ j_b \cdot i_o & j_b \cdot j_o & j_b \cdot k_o \\ k_b \cdot i_o & k_b \cdot j_o & k_b \cdot k_o \end{bmatrix}. \tag{2.7}$$

Similarly, the coordinates of $\overline{r}_p$ in the fixed frame can be expressed in terms of the coordinates in the body frame as

$$\overline{r}_p^o = R_b^o \overline{r}_p^b \tag{2.8}$$

where

$$R_b^o = \begin{bmatrix} i_o \cdot i_b & i_o \cdot j_b & i_o \cdot k_b \\ j_o \cdot i_b & j_o \cdot j_b & j_o \cdot k_b \\ k_o \cdot i_b & k_o \cdot j_b & k_o \cdot k_b \end{bmatrix}. \tag{2.9}$$

Since dot products are commutative, from equations (2.7) and (2.9), we can see that

$$R_b^o = (R_o^b)^T = (R_o^b)^{-1} \tag{2.10}$$

and

$$R_o^b = (R_b^o)^T = (R_b^o)^{-1}. \tag{2.11}$$

Then,

$$R_o^b R_b^o = (R_b^o)^T R_b^o = R_o^b (R_o^b)^T = I_3 \tag{2.12}$$

where I_3 is the 3×3 identity matrix. The transformations R_b^o and R_o^b are orthonormal transformations since all their column vectors are unit vectors in addition to being orthogonal. Thus, the transformation matrices can be used to relate the representations of the same vector in two different coordinate reference frames. Note that the transformations do not change the vector itself but only its representation. Another important thing to be noted is that the rows of R_b^o are the direction cosines of i_o, j_o, and k_o, respectively, in the body coordinate frame.

Matrices R_o^b and R_b^o can also be interpreted as rotation matrices wherein (i_b, j_b, k_b) are orthonormal basis vectors in the final direction of the (i_o, j_o, k_o) axis system after rotations about the selected coordinate axes. The properties of rotation matrices are developed in the next section.

2.2.2 Basic Rotation Matrices

When the rotation matrix represents a change in the orientation about any one of the principal coordinates of the reference frame, X, Y, or Z, it is called as the 'basic' rotation matrix. So, if a new coordinate system, say $\mathcal{B}$ with axes $(\overline{x}, \overline{y}, \overline{z})$, is obtained by rotation α of the old coordinate system , say $\mathcal{A}$ with axes (x, y, z), about X axis, then the basic rotation matrix associated with this rotation is given by, $R_{\mathcal{B}}^{\mathcal{A}} = R_{X,\alpha}$. Similarly, the basic rotation matrices associated with rotations, β about Y axis and γ about Z axis, are given by $R_{Y,\beta}$ and $R_{Z,\gamma}$, respectively. Referring to Figures (2.2a)-(2.2c), the basic rotation matrices can be written as

$$R_{X,\alpha} = \begin{bmatrix} 1 & 0 & 0 \\ 0 & cos\alpha & -sin\alpha \\ 0 & sin\alpha & cos\alpha \end{bmatrix} \tag{2.13}$$

$$R_{Y,\beta} = \begin{bmatrix} cos\beta & 0 & sin\beta \\ 0 & 1 & 0 \\ -sin\beta & 0 & cos\beta \end{bmatrix} \tag{2.14}$$

$$R_{Z,\gamma} = \begin{bmatrix} cos\gamma & -sin\gamma & 0 \\ sin\gamma & cos\gamma & 0 \\ 0 & 0 & 1 \end{bmatrix}. \tag{2.15}$$

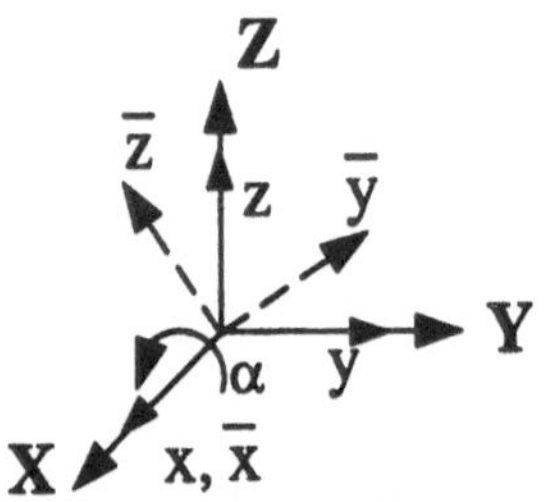

(a) X-rotation

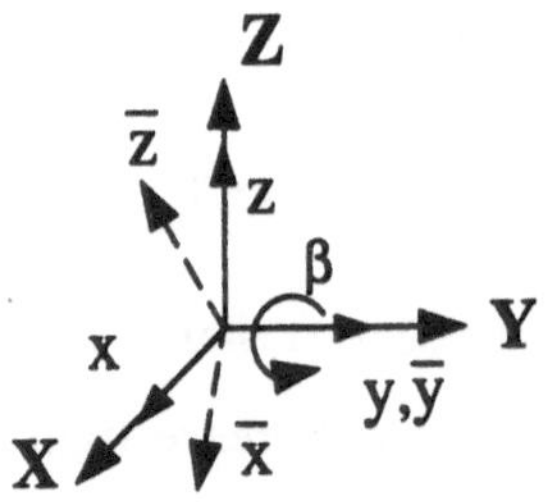

(b) Y-rotation

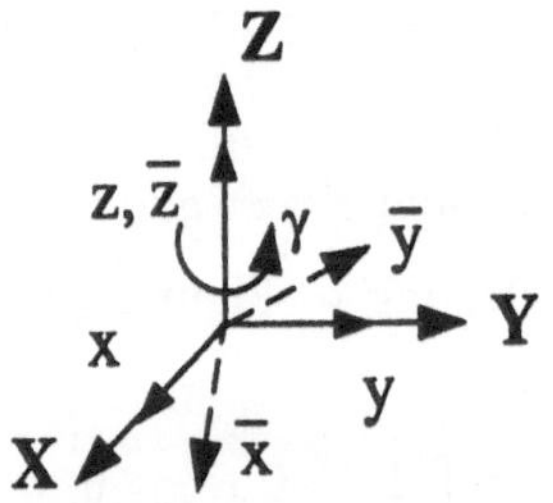

(c) Z-rotation

Figure 2.2: Basic rotations

The reason these matrices are called basic rotation matrices is because any finite arbitrary rotation can be achieved by a composition of these matrices. However, since the finite rotations are not commutative, the order of multiplication of these matrices during composition is very important.

2.2.3 Composite Rotations

As stated previously, any arbitrary finite rotation can be achieved by a composition of the basic rotation matrices, i.e., by following a sequence of basic rotations. In obtaining the composite rotation matrix there are three different possibilities. The successive rotations can take place either about the prinicipal axes of the fixed reference frame, or it can take place about the principal axes of the rotating frame itself, or a combination of both. The following procedure can be followed to obtain a composite rotation matrix

When the rotation occurs about any principal axis of a fixed reference frame, premultiply the last resultant rotation matrix by the corresponding basic roation matrix and, when the rotation occurs about any principal axis of the rotating reference frame itself, postmultiply the last resultant rotation matrix by corresponding basic rotation matrix. Following example can help illustrate these operations more clearly.

Let us suppose that the two axes systems, $OXYZ$ (fixed) and $oxyz$ (rotating), are initially coincident. Then the initial rotation matrix will just be an identity matrix, I_3. Now suppose that $oxyz$ undergoes the following sequence of rotations. First, it rotates about OX-axis through angle α, and then rotates about ox-axis (i.e. x-axis of rotating frame) through angle ϕ. The composite rotation matrix is then given by

$$R = R_{X,\alpha} I_3 R_{x,\phi} = R_{X,\alpha} R_{x,\phi} = R_{X,(\alpha+\phi)} = R_{x,(\alpha+\phi)} \tag{2.16}$$

$$\begin{aligned} R &= \begin{bmatrix} 1 & 0 & 0 \\ 0 & cos\alpha & -sin\alpha \\ 0 & sin\alpha & cos\alpha \end{bmatrix} \begin{bmatrix} 1 & 0 & 0 \\ 0 & cos\phi & -sin\phi \\ 0 & sin\phi & cos\phi \end{bmatrix} \\ &= \begin{bmatrix} 1 & 0 & 0 \\ 0 & cos(\alpha+\phi) & -sin(\alpha+\phi) \\ 0 & sin(\alpha+\phi) & cos(\alpha+\phi) \end{bmatrix}. \end{aligned} \tag{2.17}$$

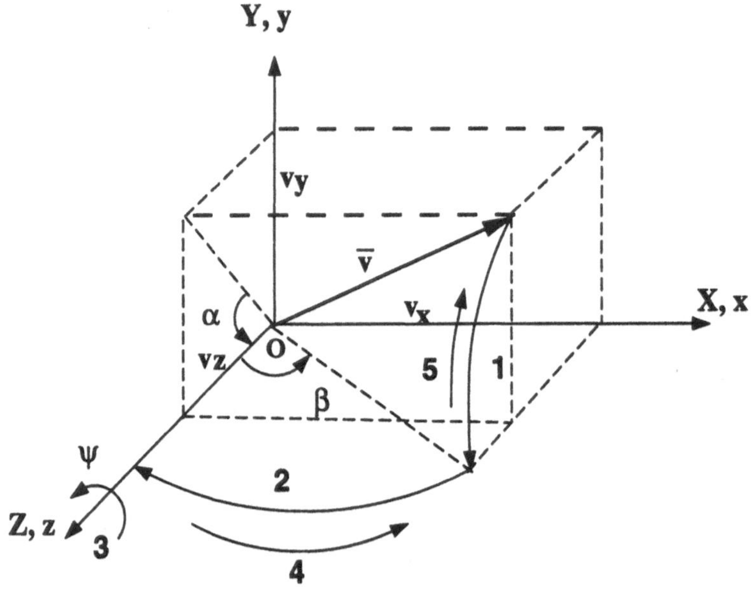

Figure 2.3: Rotation about an arbitrary axis

2.2.4 Rotation About an Arbitrary Axis

In many cases, the rotations of the body-fixed coordinate system (x, y, z) take place about an axis other than the principal axes of the fixed frame or the rotating frame. In the case when the rotation takes place about an arbitrary axis, the rotation matrix can be obtained as follows [Fu.87]. Referring to Figure 2.3, let $\overline{v} = \{v_X, v_Y, v_Z\}^T$ be the unit vector along the axis of rotation and ψ be the angle through which the rotation takes place. Now obtain the rotation matrix for following sequence of operations. Rotate $\overline{v}$ about OX by angle α which will bring vector $\overline{v}$ in the XZ plane. Then rotating about OY axis by $-\beta$ will align $\overline{v}$ vector with OZ axis. Then rotate about OZ or $\overline{v}$ by angle ψ and then reverse the rotation order to bring vector $\overline{v}$ back to its original position. This sequence of rotation will lead to the following composition of basic rotation matrices.

$$R = R_{X,-\alpha} R_{Y,\beta} R_{Z,\psi} R_{Y,-\beta} R_{X,\alpha}. \tag{2.18}$$

Noting that,

$$sin\alpha = \frac{v_Y}{\sqrt{v_Y^2 + v_Z^2}} \quad cos\alpha = \frac{v_Z}{\sqrt{v_Y^2 + v_Z^2}} \tag{2.19}$$

$$sin\beta = v_X \quad cos\beta = \sqrt{v_Y^2 + v_Z^2} \tag{2.20}$$

the rotation matrix can be rewritten as [Fu.87]

$$R = \begin{bmatrix} v_X^2 \overline{C}\psi + C\psi & v_X v_Y \overline{C}\psi - v_Z S\psi & v_X v_Z \overline{C}\psi + v_Y S\psi \\ v_X v_Y \overline{C}\psi + v_Z S\psi & v_Y^2 \overline{C}\psi + C\psi & v_Y v_Z \overline{C}\psi - v_X S\psi \\ v_X v_Z \overline{C}\psi - v_Y S\psi & v_Y v_Z \overline{C}\psi + v_X S\psi & v_Z^2 \overline{C}\psi + C\psi \end{bmatrix} \tag{2.21}$$

where $C\psi$ and $S\psi$ denote the terms $cos(\psi)$ and $sin(\psi)$, respectively, and $\overline{C}\psi = (1 - cos(\psi))$.

In mathematical notation, these 3×3 rotational matrices are said to belong to $\mathcal{SO}(3)$ space. The notation $\mathcal{SO}(3)$ stands for Special Orthogonal group of order 3 [Spo.89].

2.2.5 Properties of Rotational Matrices

The rotational matrices have some special properties which play an important role in the mathematical modeling of the system. These properties are listed below.

1) As shown previously, $R^T = R^{-1}$, i.e., $RR^T = I_3$.

2) The columns of R represent the unit vectors along the principal axes of the rotated coordinate frame with respect to the reference frame unit vectors.

3) Since $R^T = R^{-1}$ the rows of rotation matrix represent the unit vectors along principal axes of reference frame with respect to rotating frame.

4) Any row (column) of rotation matrix is orthonormal to any other row (column). This is the direct consequence of properties 1 and 2.

5) If $\overline{a}$, $\overline{b} \in \Re^3$, where $\Re^3$ is 3−dimensional Euclidean space, then $R(\overline{a} \times \overline{b}) = R\overline{a} \times R\overline{b}$ where symbol $\times$ denotes vector cross product. (Note that this equality is valid only for orthogonal rotational matrices)

2.2.6 Skew-Symmetric Matrices and Cross Product Operator

A skew-symmetric matrix, $\mathcal{S}$, has the property: $s_{ii} = 0$ and $s_{ij} = -s_{ji}$ for $i \neq j$ (where s_{ij} denotes the ij-th element of $\mathcal{S}$). Then, an immediate consequence of this property is given by: $\mathcal{S} + \mathcal{S}^T = 0$. These matrices play an important role in the computation of vector-matrix operations involving vectors belonging to $\Re^3$ space and matrices belonging to $\mathcal{RO}(3)$ space. To illustrate, consider a vector cross product $\overline{a} \times \overline{b}$ which can be written in terms of vector-matrix multiplication as $\mathcal{S}(\overline{a})\overline{b}$, where $\mathcal{S}(\cdot)$ is referred as the 'cross product operator matrix', and is given by

$$\mathcal{S}(\overline{a}) = \begin{bmatrix} 0 & -a_z & a_y \\ a_z & 0 & -a_x \\ -a_y & a_x & 0 \end{bmatrix}. \tag{2.22}$$

This cross product operator matrix has some important properties which are discussed below. For the sake of simplicity, we shall interchangeably use the following simplified notation in the remainder of the chapter.

$$\widetilde{(\cdot)} = \mathcal{S}(\cdot). \tag{2.23}$$

2.2.7 Properties of Cross Product Operator Matrix

1) Using property (5) of rotational matrices and the definition of $\mathcal{S}(\cdot)$, it can be shown [Spo.89] that for any vector, $\overline{v} \in \Re^3$ and $R \in \mathcal{SO}(3)$,

$$R(\tilde{\overline{v}})R^T = (\widetilde{R\overline{v}}). \tag{2.24}$$

2) If $R_{\overline{v},\theta}$ represents the rotation about an axis aligned with vector $\overline{v}$ by angle θ then the derivative of R with respect to θ is given by [Spo.89]

$$\frac{dR_{\overline{v},\theta}}{d\theta} = \mathcal{S}(\overline{v})R_{\overline{v},\theta} = (\tilde{\overline{v}})R_{\overline{v},\theta}. \tag{2.25}$$

3) As an obvious consequence of the cross product property, we obtain another property: $(\tilde{\overline{v}}_1)\overline{v}_2 = -(\tilde{\overline{v}}_2)\overline{v}_1$, where $\overline{v}_1$ and $\overline{v}_2$ are 3-vectors.

4) The cross product operator matrix can also be effectively used in the evaluation of the time derivative of a unit vector. For example, if $\overline{u}$ is the unit vector then

$$\frac{d}{dt}\overline{u} = \omega \times \overline{u} = \tilde{\omega}\overline{u} \tag{2.26}$$

where ω is the absolute angular velocity of the unit vector and $\tilde{\omega} = \mathcal{S}(\omega)$.

2.3 Model Derivation

The objective of this section is to derive the closed-form equations of motion for a generic nonlinear, multibody, flexible spacecraft.

2.3.1 Modeling Considerations

The focus configuration of a generic spacecraft under consideration has a branched geometry with a relatively large central body. The system under consideration is schematically represented by the configuration shown in Figure 2.4. It is assumed that all bodies in the system are flexible. The deformations in the bodies are assumed to be due to elastic motions only; however, any other deformations such as, due to thermal effects, can also be modeled if required. The system model under consideration has cluster-configuration. It consists of a central body attached to which are various appendage-bodies to form a branched geometry. For the purpose of derivation the following notations will be used. Let each body be denoted by B_{ij} where, the first subscript indicates the branch number the body belongs to and the second subscript indicates the body number in that particular branch. Since the number and the locations of various bodies are arbitrary the system configuration is very general.

2.3.2 Coordinate Systems

Referring to Figure 2.5, (X_I, Y_I, Z_I) is the inertial coordinate system; (X_{cm}, Y_{cm}, Z_{cm}) is the coordinate system with the origin fixed at the center of mass of the entire spacecraft and is aligned with the inertial frame; (X_c, Y_c, Z_c) is the coordinate frame attached to the central body with the origin attached to the

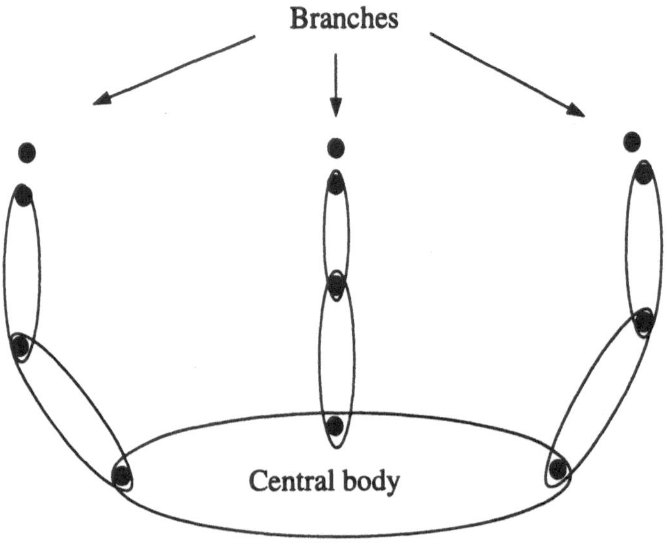

Figure 2.4: Schematic of a multibody spacecraft

center of mass, and (X_{ij}, Y_{ij}, Z_{ij}) represents local coordinate system attached to the ij-th body with the origin located at the point of connection between $i(j-1)$-th body and ij-th body. The motion of each local coordinate system origin, O_{ij}, is defined with respect to the previous local coordinate frame in the chain. In order to derive dynamical equations, it is necessary to obtain the expressions for the kinematic quantities, the position, the velocity, and the acceleration of the spacecraft.

2.3.3 Kinematics of a Spacecraft

Consider a spacecraft with its instantaneous center of mass located at O_{cm} (Figure 2.5). Then, the vector, $\overline{r}_{cm}$, determines the inertial position of the spacecraft. In order to decouple the translational motion of the entire spacecraft from the rotational motion about its center of mass, for the reasons that will be apparent later, the orientation of the frame (X_{cm}, Y_{cm}, Z_{cm}) is assumed to be same as (X_I, Y_I, Z_I), i.e. the rotation matrix, R^I_{cm}, will be an identity matrix. The vector $\overline{r}_{cm}$ can be expressed in terms of orbital elements [Ng.88].

Vector $\overline{a}_c$ represents the position vector of the center of mass of the central body due to rigid displacement only and $\overline{\rho}_c$ represents the displacement of the center of mass due to elastic motion. Vector $\overline{r}_c$ then represents the vector sum of these two vectors, i.e., $\overline{r}_c = \overline{a}_c + \overline{\rho}_c$. In addition to the translational motion of the entire spacecraft, the spacecraft can undergo the rotational motion about the combined center of mass. This motion can be characterized by the rotational transformation, R_c^{cm}, between (X_{cm}, Y_{cm}, Z_{cm}) and (X_c, Y_c, Z_c). Since, (X_{cm}, Y_{cm}, Z_{cm}) and (X_I, Y_I, Z_I) are aligned, R_c^{cm} describes the orientation of the spacecraft with respect to the inertial frame also. R_c^{cm} is generally described using Euler rotations. Other rigid body degrees of freedom arise from the interconnections between different bodies of the spacecraft, each of which can be described by the transformations of the type, $R_{ij}^{i(j-1)}$, between any two consecutive body frames in the chain. O_{ij} represents the origin of the ij-th body frame (attached to ij-th body) and its position with respect to $i(j-1)$-th body frame (attached to $i(j-1)$-th body) is defined by vector $\overline{s}_{ij}$. Also, each $\overline{s}_{ij}$ is the vector sum of $\overline{a}_{ij}$ and $\overline{\rho}_{ij}$, where $\overline{a}_{ij}$ and $\overline{\rho}_{ij}$ represent rigid body and elastic displacements of ij-th frame, respectively.

Having established the notations, we shall obtain the equations for position, velocity, and acceleration for a representative particle mass dm in the ij-th body, i.e., the j-th body in the i-th branch.

Referring to Figure 2.6, the position vector of a particle mass dm in ij-th body, in the local reference frame (i.e, ij-th frame), is given by

$$\overline{u}_{dm}^{ij} = \overline{d}_{dm}^{ij} + \overline{\rho}_{dm}^{ij} \tag{2.27}$$

where $\overline{d}_{dm}^{ij}$ and $\overline{\rho}_{dm}^{ij}$ represent rigid and elastic displacements of mass dm in ij-th frame, respectively. The position vector of dm with respect to $i(j-1)$-th frame is then given by

$$\overline{u}_{dm}^{i(j-1)} = \overline{s}_{ij} + R_{ij}^{i(j-1)} \overline{u}_{dm}^{ij}. \tag{2.28}$$

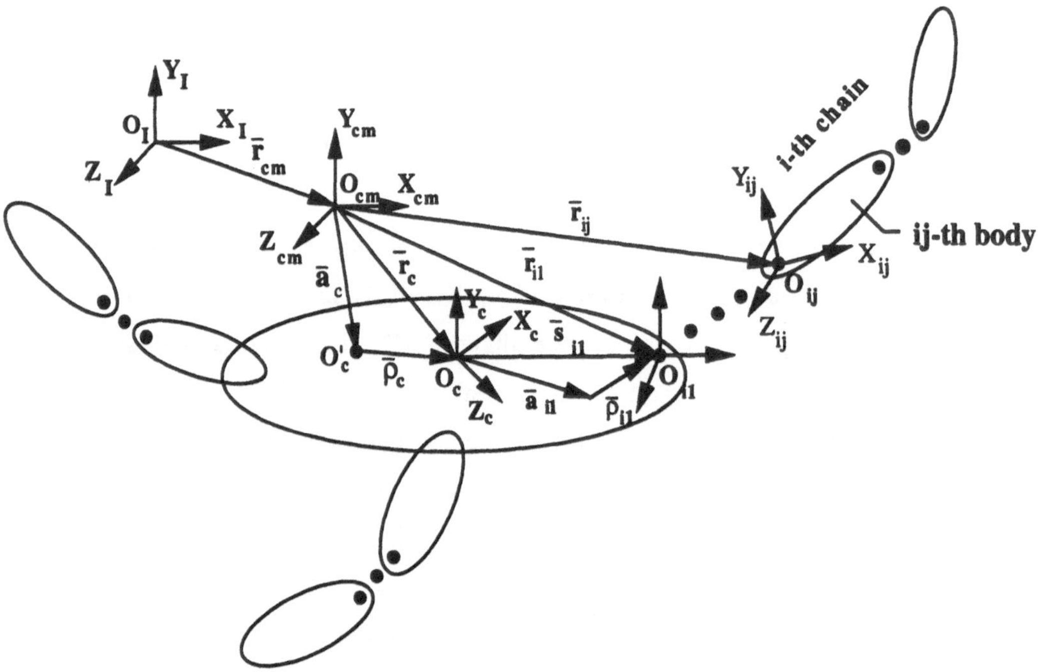

Figure 2.5: Coordinate systems

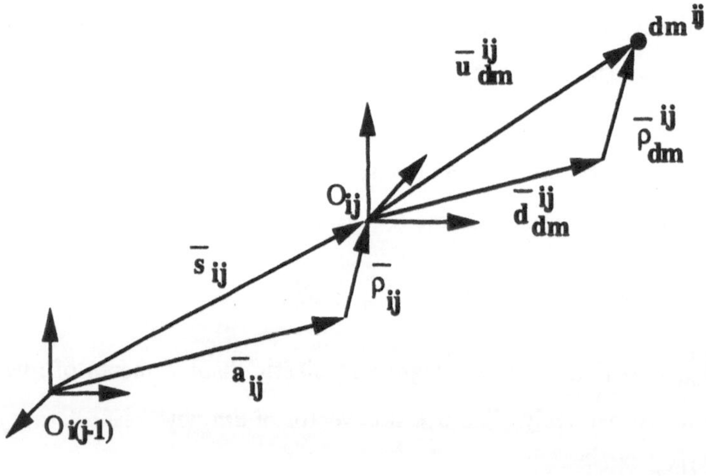

Figure 2.6: Position vector of particle mass dm in ij-th body

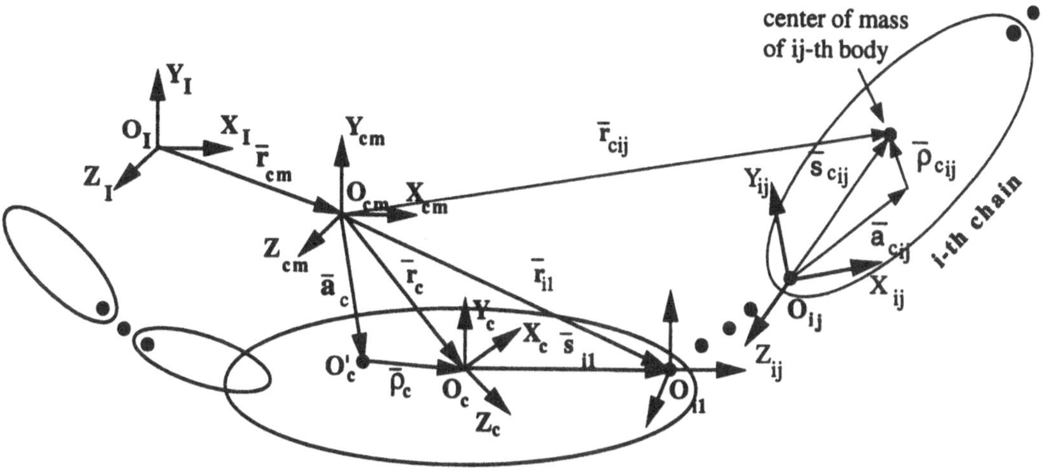

Figure 2.7: Position vector of the center of mass of ij-th body

Finally, the position vector of dm in the inertial frame of reference is given by

$$\overline{r}^{ij}_{dm} = \overline{r}_{cm} + R^I_{cm}\overline{r}_c + R^I_c\overline{s}_{i1} + \left(\sum_{k=1}^{k=j-1} R^I_{ik}\overline{s}_{i(k+1)}\right) + R^I_{ij}\overline{u}^{ij}_{dm} \tag{2.29}$$

where R^I_{ij} is given by

$$R^I_{ij} = R^I_{cm}R^{cm}_c R^c_{i1}\cdots R^{i(j-1)}_{ij}. \tag{2.30}$$

For the rest of the development here, it will be assumed that each vector is represented in its local frame of reference unless otherwise stated.

The velocity of the particle mass dm is given by taking the time derivative of equation (2.29).

$$\begin{aligned}\overline{v}^{ij}_{dm} = \dot{\overline{r}}^{ij}_{dm} &= \dot{\overline{r}}_{cm} + \frac{d}{dt}(R^I_{cm}\overline{r}_c) + \frac{d}{dt}\left(R^I_c\overline{s}_{i1}\right) + \frac{d}{dt}\left(\sum_{k=1}^{k=j-1} R^I_{ik}\overline{s}_{i(k+1)}\right)\\ &+ \frac{d}{dt}\left(R^I_{ij}\overline{u}^{ij}_{dm}\right)\\ &= \dot{\overline{r}}_{cm} + \dot{R}^I_{cm}\overline{r}_c + R^I_{cm}\dot{\overline{r}}_c + \dot{R}^I_c\overline{s}_{i1} + R^I_c\dot{\overline{s}}_{i1}\end{aligned}$$

$$
\begin{aligned}
&+ \sum_{k=1}^{k=j-1} (\dot{R}^I_{ik}\bar{s}_{i(k+1)}) + \sum_{k=1}^{k=j-1} (R^I_{ik}\dot{\bar{s}}_{i(k+1)}) \\
&+ \dot{R}^I_{ij}\bar{u}^{ij}_{dm} + R^I_{ij}\dot{\bar{u}}^{ij}_{dm}. \qquad (2.31)
\end{aligned}
$$

The term $\dot{\bar{r}}_{cm}$ in equation (2.31) represents pure translational velocity of the combined center of mass.

In practice, for many spacecraft applications, it suffices for control engineers to use only the rotational dynamic model of the spacecraft since the translational motion is controlled by separate thrusters when periodic reboosting is necessary to correct orbit decay. In view of this, our interest here will be to obtain only the rotational dynamic model of the spacecraft. As a result, it will be assumed that there are no external resultant forces acting on the system, i.e., $\sum F_{ext} = 0$. Under this condition, the principle of conservation of linear momentum applies and the total linear moment of inertia of the system will be conserved. Then, assuming zero initial condition for the translational velocity of the spacecraft (i.e., $\dot{\bar{r}}_{cm} = 0$), the conservation of linear momentum yields

$$
m_c\dot{\bar{r}}_c + \sum_i \sum_j m_{ij}\dot{\bar{r}}_{cij} = 0 \qquad (2.32)
$$

where m_c and m_{ij} are masses of the central body and ij-th body in the system, respectively, and $\bar{r}_{cij}$ (as shown in Figure 2.7) is the position vector of the center of mass of the ij-th body with respect to O_{cm}.

$$
\begin{aligned}
m_c\dot{\bar{r}}_c &= -\sum_i \sum_j m_{ij}\dot{\bar{r}}_{cij} \\
&= -\sum_i \sum_j m_{ij}(\dot{\bar{r}}_c + \sum_{k=1}^{j} \dot{\bar{s}}_{ik} + \dot{\bar{s}}_{cij}) \\
&= -\sum_i \sum_j m_{ij}\dot{\bar{r}}_c - \left[\sum_i \sum_j m_{ij}(\sum_{k=1}^{j} \dot{\bar{s}}_{ik} + \dot{\bar{s}}_{cij})\right] \\
\text{i.e.,} \quad (\sum_i \sum_j m_{ij} + m_c)\dot{\bar{r}}_c &= -\sum_i \sum_j m_{ij}(\sum_{k=1}^{j} \dot{\bar{s}}_{ik} + \dot{\bar{s}}_{cij}) \\
\dot{\bar{r}}_c &= -\frac{1}{M}\sum_i \sum_j m_{ij}(\sum_{k=1}^{j} \dot{\bar{s}}_{ik} + \dot{\bar{s}}_{cij}) \qquad (2.33)
\end{aligned}
$$

where $\bar{s}_{cij}$ represents the position vector of the center of mass of ij-th body in its own body frame and $M = (\sum_i \sum_j m_{ij} + m_c)$ is the total mass of the spacecraft. Noting that $\dot{\bar{r}}_{cm} = 0$, $\dot{R}^I_{cm} = 0$, and substituting equation (2.33) in equation (2.31) yields

$$\begin{aligned}
\bar{v}^{ij}_{dm} = \dot{\bar{r}}^{ij}_{dm} &= -\frac{1}{M} R^I_{cm} \left(\sum_i \sum_j m_{ij} (\sum_{k=1}^{j} \dot{\bar{s}}_{ik} + \dot{\bar{s}}_{cij}) \right) + \dot{R}^I_c \bar{s}_{i1} + R^I_c \dot{\bar{s}}_{i1} \\
&+ \sum_{k=1}^{k=j-1} (\dot{R}^I_{ik} \bar{s}_{i(k+1)}) + \sum_{k=1}^{k=j-1} (R^I_{ik} \dot{\bar{s}}_{i(k+1)}) \\
&+ \dot{R}^I_{ij} \bar{u}^{ij}_{dm} + R^I_{ij} \dot{\bar{u}}^{ij}_{dm}.
\end{aligned} \tag{2.34}$$

Now using the properties of the cross product operator (section 2.2.7), various time derivatives appearing in equation (2.34) can be evaluated as follows:

$$\dot{R}^I_c \bar{s}_{i1} = \dot{R}^{cm}_c \bar{s}_{i1} = \tilde{\omega}_c R^{cm}_c \bar{s}_{i1} = -\mathcal{S}(R^{cm}_c \bar{s}_{i1}) \bar{\omega}_c \tag{2.35}$$

$$\begin{aligned}
\dot{\bar{s}}_{i1} &= \dot{\bar{a}}_{i1} + \dot{\bar{\rho}}_{i1} = \tilde{\bar{\omega}}_c (\bar{a}_{i1} + \sum_k \Phi^{O_{i1}}_{ck} q_{ck}) + \sum_k \Phi^{O_{i1}}_{ck} \dot{q}_{ck} \\
&= -\mathcal{S}(\bar{a}_{i1} + \sum_k \Phi^{O_{i1}}_{ck} q_{ck}) \bar{\omega}_c + \sum_k \Phi^{O_{i1}}_{ck} \dot{q}_{ck}
\end{aligned} \tag{2.36}$$

where, $\bar{\rho}_{i1} = \sum_k \Phi^{O_{i1}}_{ck} q_{ck}$, and $\Phi^{O_{i1}}_{ck}$ is the $(3 \times k)$ mode shape matrix of the central body at O_{i1}, and q_{ck} are the generalized coordinates (modal amplitudes). Similarly, for $j > 1$,

$$\begin{aligned}
\dot{\bar{s}}_{ij} &= \dot{\bar{a}}_{ij} + \dot{\bar{\rho}}_{ij} = \tilde{\bar{\omega}}_{i(j-1)} (\bar{a}_{ij} + \sum_k \Phi^{O_{ij}}_{i(j-1)k} q_{i(j-1)k}) + \sum_k \Phi^{O_{ij}}_{i(j-1)k} \dot{q}_{i(j-1)k} \\
&= -\mathcal{S}\left(\bar{a}_{ij} + \sum_k \Phi^{O_{ij}}_{i(j-1)k} q_{i(j-1)k} \right) \bar{\omega}_{i(j-1)} + \sum_k \Phi^{O_{ij}}_{i(j-1)k} \dot{q}_{i(j-1)k}
\end{aligned} \tag{2.37}$$

where, $\dot{\bar{\rho}}_{ij} = \sum_k \Phi^{O_{ij}}_{i(j-1)k} \dot{q}_{i(j-1)k}$, $\Phi^{O_{ij}}_{i(j-1)k}$ is the k-th mode shape matrix at O_{ij}, and $q_{i(j-1)k}$ are the generalized coordinates for $i(j-1)$-th body. Also,

$$\begin{aligned}
\dot{\bar{s}}_{cij} &= \dot{\bar{a}}_{cij} + \dot{\bar{\rho}}_{cij} = \tilde{\bar{\omega}}_{ij} \left(\bar{a}_{cij} + \sum_k \Phi^c_{ijk} q_{ijk} \right) + \sum_k \Phi^c_{ijk} \dot{q}_{ijk} \\
&= -\mathcal{S}\left(\bar{a}_{cij} + \sum_k \Phi^c_{ijk} q_{ijk} \right) \bar{\omega}_{ij} + \sum_k \Phi^c_{ijk} \dot{q}_{ijk}.
\end{aligned} \tag{2.38}$$

where Φ^{c}_{ijk} is the mode-shape matrix at the center of mass location of ij-th body.

$$\begin{aligned}
\dot{R}^{I}_{ik}\overline{s}_{i(k+1)} &= \dot{R}^{cm}_{c}R^{c}_{i1}\cdots R^{i(k-1)}_{ik}\overline{s}_{i(k+1)} + R^{cm}_{c}\dot{R}^{c}_{i1}\cdots R^{i(k-1)}_{ik}\overline{s}_{i(k+1)} \\
&+ \cdots - R^{cm}_{c}R^{c}_{i1}\cdots \dot{R}^{i(k-1)}_{ik}\overline{s}_{i(k+1)} \\
&= -\mathcal{S}\Big(R^{cm}_{c}R^{c}_{i1}\cdots R^{i(k-1)}_{ik}\overline{s}_{i(k+1)}\Big)\overline{\omega}_{c} - R^{cm}_{c}\mathcal{S}\Big(R^{c}_{i1}\cdots R^{i(k-1)}_{ik}\overline{s}_{i(k+1)}\Big)\overline{\omega}_{i1} \\
&+ \cdots + R^{cm}_{c}R^{c}_{i1}\cdots R^{i(k-2)}_{i(k-1)}\mathcal{S}\Big(R^{i(k-1)}_{ik}\overline{s}_{i(k+1)}\Big)\overline{\omega}_{ik} \qquad (2.39) \\
\dot{R}^{I}_{ij}\overline{u}^{ij}_{dm} &= -\mathcal{S}\Big(R^{cm}_{c}R^{c}_{i1}\cdots R^{i(j-1)}_{ij}\overline{u}^{ij}_{dm}\Big)\overline{\omega}_{c} - R^{cm}_{c}\mathcal{S}\Big(R^{c}_{i1}\cdots R^{i(j-1)}_{ij}\overline{u}^{ij}_{dm}\Big)\overline{\omega}_{i1} \\
&+ \cdots + R^{I}_{cm}R^{cm}_{c}R^{c}_{i1}\cdots R^{i(j-2)}_{i(j-1)}\mathcal{S}\Big(R^{i(j-1)}_{ij}\overline{u}^{ij}_{dm}\Big)\overline{\omega}_{ij} \qquad (2.40) \\
R^{I}_{ij}\dot{\overline{u}}^{ij}_{dm} &= R^{I}_{ij}(\dot{\overline{d}}^{ij}_{dm} + \dot{\overline{\rho}}^{ij}_{dm}) \\
&= R^{I}_{ij}(\tilde{\overline{\omega}}_{ij}(\overline{d}^{ij}_{dm} + \sum_{k}\Phi^{dm}_{ijk}q_{ijk}) + \sum_{k}\Phi^{dm}_{ijk}\dot{q}_{ijk}) \\
&= -R^{I}_{ij}\mathcal{S}\Big(\overline{d}^{ij}_{dm} + \sum_{k}\Phi^{dm}_{ijk}q_{ijk}\Big)\overline{\omega}_{ij} + R^{I}_{ij}\sum_{k}\Phi^{dm}_{ijk}\dot{q}_{ijk}. \qquad (2.41)
\end{aligned}$$

Substituting equations (2.35)-(2.41) in equation (2.34), and noting that $R^{I}_{cm} = I_3$, we get

$$\begin{aligned}
\overline{v}^{ij}_{dm} &= -\frac{1}{M}\sum_{i}\sum_{j}m_{ij}\sum_{k=1}^{j}\Big[-\mathcal{S}\Big(\overline{a}_{ik} + \sum_{l}\Phi^{O_{ik}}_{i(k-1)l}q_{i(k-1)l}\Big)\overline{\omega}_{i(k-1)} \\
&+ \sum_{l}\Phi^{O_{ik}}_{i(k-1)l}\dot{q}_{i(k-1)l}\Big] - \frac{1}{M}\sum_{i}\sum_{j}m_{ij}\Big[-\mathcal{S}\Big(\overline{a}_{cij} + \sum_{l}\Phi^{cij}_{ijl}q_{ijl}\Big)\overline{\omega}_{ij} \\
&+ \sum_{l}\Phi^{cij}_{ijl}\dot{q}_{ijl}\Big] - \mathcal{S}(R^{cm}_{c}\overline{s}_{i1})\overline{\omega}_{c} - R^{I}_{c}\mathcal{S}\Big(\overline{a}_{i1} + \sum_{l}\Phi^{O_{i1}}_{cl}q_{cl}\Big)\overline{\omega}_{c} + R^{I}_{c}\sum_{l}\Phi^{O_{i1}}_{cl}\dot{q}_{cl} \\
&+ \sum_{k=1}^{j-1}\Big[-\mathcal{S}\Big(R^{cm}_{c}\cdots R^{i(k-1)}_{ik}\overline{s}_{i(k+1)}\Big)\overline{\omega}_{c} - R^{cm}_{c}\mathcal{S}(R^{c}_{ik}\overline{s}_{i(k+1)})\overline{\omega}_{i1} - \cdots \\
&- R^{cm}_{i(k-1)}\mathcal{S}(R^{i(k-1)}_{ik}\overline{s}_{i(k+1)})\overline{\omega}_{ik}\Big] \\
&+ \sum_{k=1}^{j-1}\Big[-R^{I}_{ik}\mathcal{S}\Big(\overline{a}_{i(k+1)} + \sum_{l}\Phi^{O_{i(k+1)}}_{ikl}q_{ikl}\Big)\overline{\omega}_{ik} + R^{I}_{ik}\sum_{l}\Phi^{O_{i(k+1)}}_{ikl}\dot{q}_{ikl}\Big] \\
&- \mathcal{S}(R^{cm}_{ij}\overline{u}^{ij}_{dm})\overline{\omega}_{c} - R^{cm}_{c}\mathcal{S}(R^{c}_{ij}\overline{u}^{ij}_{dm})\overline{\omega}_{i1} - \cdots - R^{I}_{i(j-1)}\mathcal{S}(R^{i(j-1)}_{ij}\overline{u}^{ij}_{dm})\overline{\omega}_{ij} \\
&- R^{I}_{ij}\mathcal{S}\Big(\overline{d}^{ij}_{dm} + \sum_{l}\Phi^{dm}_{ijl}q_{ijl}\Big)\overline{\omega}_{ij} + R^{I}_{ij}\sum_{l}\Phi^{dm}_{ijl}\dot{q}_{ijl}. \qquad (2.42)
\end{aligned}$$

Simplifying and regrouping equation (2.42) gives

$$\begin{aligned}
\overline{v}^{ij}_{dm} &= \Bigg\{-\frac{1}{M}\sum_i\sum_j m_{ij}\Big[-\mathcal{S}\Big(\overline{a}_{i1}+\sum_l \Phi^{O_{i1}}_{icl}q_{icl}\Big)\Big] - \mathcal{S}(R^{cm}_c\overline{s}_{i1}) \\
&- R^I_c\mathcal{S}\Big(\overline{a}_{i1}+\sum_l \Phi^{O_{i1}}_{cl}q_{cl}\Big) - \sum_{k=1}^{j-1}\mathcal{S}(R^{cm}_{ik}\overline{s}_{i(k+1)}) - \mathcal{S}(R^{cm}_{ij}\overline{u}^{ij}_{dm})\Bigg\}\overline{\omega}_c \\
&+ \Bigg\{\frac{1}{M}\sum_i\sum_j m_{ij}\Big[\mathcal{S}\Big(\overline{a}_{i2}+\sum_l \Phi^{O_{i2}}_{i1l}q_{i1l}\Big)\Big] \\
&+ \frac{1}{M}\sum_i m_{i1}\Big[\mathcal{S}\Big(\overline{a}_{ci1}+\sum_l \Phi^{O_{ci1}}_{i1l}q_{i1l}\Big)\Big] - \sum_{k=1}^{j-1} R^{cm}_c\mathcal{S}(R^c_{ik}\overline{s}_{i(k+1)}) \\
&- \sum_{k=1}^{j-1} R^I_{i1}\mathcal{S}(\overline{a}_{i2}+\sum_l \Phi^{O_{i2}}_{i1l}q_{i1l}) - R^{cm}_c\mathcal{S}(R^c_{ij}\overline{u}^{ij}_{dm})\Bigg\}\overline{\omega}_{i1} - \cdots \\
&+ \Bigg[\frac{1}{M}\sum_i\sum_j m_{ij}\Big(\mathcal{S}(\overline{a}_{cij}+\sum_l \Phi^{cij}_{ijl})\Big) - R^I_{i(j-1)}\mathcal{S}(R^{i(j-1)}_{ij}\overline{u}^{ij}_{dm}) \\
&- R^I_{ij}\mathcal{S}(\overline{d}^{ij}_{dm}+\sum_l \Phi^{dm}_{ijl}q_{ijl})\Bigg]\overline{\omega}_{ij} \\
&- \frac{1}{M}\sum_i\sum_j m_{ij}\Big[\sum_{k=1}^{j}\Big(\sum_l \Phi^{O_{ik}}_{i(k-1)l}\dot{q}_{i(k-1)l}\Big) + \sum_l \Phi^{cij}_{ijl}\dot{q}_{ijl}\Big] \\
&+ R^I_c\sum_l \Phi^{O_{i1}}_{cl}\dot{q}_{cl} + \sum_{k=1}^{j-1}\Big(R^I_{ik}\sum_l \Phi^{O_{i(k+1)}}_{ikl}\dot{q}_{ikl}\Big) \\
&+ R^I_{ij}\sum_l \Phi^{dm}_{ijl}\dot{q}_{ijl}.
\end{aligned} \tag{2.43}$$

Equation (2.43) can be rewritten in a compact form as

$$\overline{v}^{ij}_{dm} = \overline{v}^{rot}_{dm} = N\dot{p} \tag{2.44}$$

where $\overline{v}^{rot}_{dm}$ indicates that the contributions to the velocity vector of particle mass dm are from the rotational-plus-flexible motion of the spacecraft. In equation (2.44), N is $3\times n$ matrix, n is the total number of rotational-plus-flexible degrees of freedom, and $\dot{p}$ is $n\times 1$ vector of generalized velocities corresponding to these degrees of freedom. Matrix N has the following form:

$$N = [N_{rigid} \quad N_{flex}] \tag{2.45}$$

where, N_{rigid} is $3\times 3(n_c m+1)$ matrix related to the rigid degrees of freedom and N_{flex} is $3\times s(n_c m+1)$ matrix related to the flexible degrees of freedom. For the sake of notational simplicity, and without loss of generality, it is assumed that there are n_c chains, each of the chain structure has m bodies, and that each body has s flexible degrees of freedom. Then vector $\dot{p}$ has the form

$$\dot{p} = [\overline{\omega}_c, \overline{\omega}_{11}, \cdots \overline{\omega}_{1m}, \cdots \overline{\omega}_{ij}, \cdots \overline{\omega}_{n_c m}, \dot{q}_{c1}, \cdots \dot{q}_{cs}, \dot{q}_{111}, \cdots \dot{q}_{ijk}, \cdots \dot{q}_{n_c ms}]^T. \tag{2.46}$$

The matrices N_{rigid} and N_{flex} are given as follows.

$$N_{rigid} = [N_c, N_{i1}, \cdots, \cdots, N_{ij}, \cdots] \tag{2.47}$$

where

$$\begin{aligned} N_c &= \text{coefficient matrix of } \overline{\omega}_c \text{ in equation (2.43)} \\ N_{i1} &= \text{coefficient matrix of } \overline{\omega}_{i1} \text{ in equation (2.43)} \\ &\cdots \\ N_{ij} &= \text{coefficient matrix of } \overline{\omega}_{ij} \text{ in equation (2.43)} \end{aligned} \tag{2.48}$$

and

$$N_{flex} = [N_{qc} \cdots N_{q11} \cdots N_{qij} \cdots] \tag{2.49}$$

where

$$\begin{aligned} N_{qc} &= (3 \times s) \text{ coefficient matrix related to the flexural coordinates} \\ &\quad \text{of the central body} \\ &\cdots \\ N_{qij} &= (3 \times s) \text{ coefficient matrix related to the flexural coordinates} \\ &\quad \text{of the } ij\text{-th body.} \end{aligned} \tag{2.50}$$

Having obtained the expression for velocity, the kinetic energy for the particle mass dm is given by

$$dT_{dm} = \frac{1}{2} \overline{v}_{dm}^{rot\,T} \overline{v}_{dm}^{rot} dm. \tag{2.51}$$

The kinetic energy for the entire spacecraft can be obtained by integrating equation (2.51), i.e.,

$$T = \int_{\Omega} dT_{dm} = \frac{1}{2}\int_{\Omega} \overline{v}_{dm}^{rot^T} \overline{v}_{dm}^{rot} \mu d\Omega \tag{2.52}$$

where, μ is the mass density, $\overline{v}_{dm}^{rot}$ is as given in equation (2.43), and Ω denotes the spatial domain of integration. Substituting equation (2.51) into equation (2.52), we get

$$\begin{aligned} T &= \frac{1}{2}\int_{\Omega} (N\dot{p})^T N\dot{p}\mu d\Omega \\ &= \frac{1}{2}\int_{\Omega} \dot{p}^T (N^T N)\dot{p}\mu d\Omega. \end{aligned} \tag{2.53}$$

Equation (2.53) can be rewritten in a compact form as

$$T = \frac{1}{2}\dot{p}^T M(p)\dot{p} \tag{2.54}$$

where, $M(p)$ is the mass-inertia matrix of the system, given by

$$M(p) = \int_{\Omega} (N^T N)\mu d\Omega. \tag{2.55}$$

$M(p)$ is a symmetric and positive definite matrix.

It is to be noted that although the kinetic energy expression derived here assumes only rotational-plus-flexible dynamics for the spacecraft, the effects of translational motion can be easily added by modifying the term $\overline{v}_{dm}^{ij}$ in equation (2.43) to yield the kinetic energy expression which has both translational and rotational components.

2.3.4 Potential Energy

The potential energy of the system could be due to many sources, such as elastic displacement, thermal deformation, etc. The deformations due to thermal effects is not considered here, however, it can be easily included in the formulation if desired. Thus, it is assumed that the potential energy has contribution only from the elastic strain energy. Also, it assumed that the materials under consideration are isotropic in nature and that they obey Hooke's law.

For the isotropic materials obeying Hooke's law, the strain energy differential is given as

$$\delta V = \int_\Omega \sigma^T \delta\epsilon d\Omega \tag{2.56}$$

where the terms σ and $\delta\epsilon$ represent the stress and differential strain, respectively. Equation (2.56) can be rewritten as

$$\delta V = \int_\Omega \Psi d\Omega \tag{2.57}$$

where Ψ is the strain energy density and has the form

$$\Psi = \sigma_{xx}\epsilon_{xx} + \sigma_{yy}\epsilon_{yy} + \ldots + \sigma_{yz}\epsilon_{yz}. \tag{2.58}$$

Now, for materials obeying Hooke's law, following equality holds.

$$\sigma^T = E\epsilon. \tag{2.59}$$

The strain-displacement relation is given by

$$\epsilon = \mathcal{D}u \tag{2.60}$$

where, u is the general displacement vector and $\mathcal{D}$ is the differential operator defined by relations

$$\epsilon_{ij} = \frac{1}{2}\left[u_{i,j} + u_{j,i} + \sum_{k=1}^{3} u_{k,i}^{i,j}\right] \qquad \{i,j = 1,2,3\}. \tag{2.61}$$

The vector u can be expressed in terms of the modal coordinates as

$$u = \Phi q. \tag{2.62}$$

Now, from equation (2.60)

$$\delta\epsilon = \mathcal{D}\Phi\delta q. \tag{2.63}$$

Substituting equation (2.63) in equation (2.56), we get

$$\begin{aligned} \delta V &= \int_\Omega \sigma^T \delta\epsilon d\Omega \\ &= \int_\Omega \epsilon^T E\mathcal{D}\Phi\delta q d\Omega \\ &= \int_\Omega q^T (\mathcal{D}\Phi)^T E\mathcal{D}\Phi\delta q d\Omega \\ &= q^T \int_\Omega (\mathcal{D}\Phi)^T E\mathcal{D}\Phi d\Omega\delta q \\ &= q^T K\delta q \end{aligned} \tag{2.64}$$

where, K is called the stiffness matrix of the system and is given by

$$K = \int_{\Omega} (\mathcal{D}\Phi)^T E \mathcal{D}\Phi d\Omega. \tag{2.65}$$

The potential energy of the system is, then given by

$$V = \frac{1}{2} q^T K q \tag{2.66}$$

where K is given by equation (2.65).

2.3.5 Equations of Motion

As previously stated, only rotational motion of the spacecraft is considered in deriving the dynamic model of the spacecraft. However, if necessary, the effects of translational motion can be easily included in the dynamics of motion by using appropriate kinetic energy term in the Lagrangian, i.e., by using T which has both rotational and translational components.

Using equations (2.54) and (2.66) the Lagrangian of the system is defined as

$$L = T - V. \tag{2.67}$$

For the purpose of convenience, L can be rewritten in the indicial notation as

$$L = T - V = \frac{1}{2} \sum_{i,j} M_{ij} \dot{p}_i \dot{p}_j - V(q). \tag{2.68}$$

The Euler-Lagrange equations for the system can then be derived from the following equation.

$$\frac{d}{dt}\left(\frac{\partial L}{\partial \dot{p}_k}\right) - \frac{\partial L}{\partial p_k} = F_k \quad (k = 1, 2, \cdots, n) \tag{2.69}$$

where, F_k are generalized forces from non-conservative force field. Evaluating the derivatives yield

$$\frac{\partial L}{\partial \dot{p}_k} = \sum_{j} M_{kj} \dot{p}_j \tag{2.70}$$

and

$$\begin{aligned}\frac{d}{dt}\left(\frac{\partial L}{\partial \dot{p}_k}\right) &= \sum_j M_{kj}\ddot{p}_j + \sum_j \dot{M}_{kj}\dot{p}_j \\ &= \sum_j M_{kj}\ddot{p}_j + \sum_{i,j} \frac{\partial M_{kj}}{\partial p_i}\dot{p}_i\dot{p}_j. \end{aligned} \tag{2.71}$$

Also

$$\frac{\partial L}{\partial p_k} = \frac{1}{2}\sum_{i,j} \frac{\partial M_{ij}}{\partial p_k}\dot{p}_i\dot{p}_j - \frac{\partial V}{\partial p_k}. \tag{2.72}$$

Substituting equations (2.70-2.72) in equation (2.69), the Euler-Lagrange equations can be rewritten as

$$\sum_j M_{kj}\ddot{p}_j + \sum_{i,j}\left[\frac{\partial M_{kj}}{\partial p_i} - \frac{1}{2}\frac{\partial M_{ij}}{\partial p_k}\right]\dot{p}_i\dot{p}_j - \frac{\partial V}{\partial p_k} = F_k \quad (k = 1, ..., n).$$

By interchanging the order of summation and taking advantage of symmetry, it can be seen that

$$\sum_{i,j}\left(\frac{\partial M_{kj}}{\partial p_i}\right)\dot{p}_i\dot{p}_j = \frac{1}{2}\sum_{i,j}\left(\frac{\partial M_{kj}}{\partial p_i} + \frac{\partial M_{ki}}{\partial p_j}\right)\dot{p}_i\dot{p}_j. \tag{2.73}$$

Hence

$$\sum_{i,j}\left(\frac{\partial M_{kj}}{\partial p_i} - \frac{1}{2}\frac{\partial M_{ij}}{\partial p_k}\right)\dot{p}_i\dot{p}_j = \sum_{i,j}\frac{1}{2}\left(\frac{\partial M_{kj}}{\partial p_i} + \frac{\partial M_{ki}}{\partial p_j} - \frac{\partial M_{ij}}{\partial p_k}\right)\dot{p}_i\dot{p}_j. \tag{2.74}$$

The terms

$$C_{ijk} = \frac{1}{2}\left(\frac{\partial M_{kj}}{\partial p_i} + \frac{\partial M_{ki}}{\partial p_j} - \frac{\partial M_{ij}}{\partial p_k}\right) \tag{2.75}$$

are known as Christoffel symbols. Note that, for each fixed k, we have $C_{ijk} = C_{jki}$. Also

$$\frac{\partial V}{\partial p_k} = K_{kj}q_j. \tag{2.76}$$

Finally, the Euler-Lagrange equations of motion can be rewritten as

$$\sum_j M_{kj}\ddot{p}_j + \sum_{i,j} C_{ijk}\dot{p}_i\dot{p}_j + D_{kj}\dot{p} + K_{kj}p_j = F_k \qquad (k = 1, 2, ..., n) \tag{2.77}$$

where D is the inherent structural damping matrix and $D\dot{p}$ is the vector of nonconservative forces.

In equation (2.77), there are four types of terms. The first type of terms involve the second time derivative of the generalized coordinates. The second type of terms are quadratic terms in the first time derivatives of p, where the coefficients may depend on p. These terms can be further classified into two types: the terms involving the products of the type $\dot{p}^2$, called *centrifugal* terms, and those involving the products of the type $\dot{p}_i\dot{p}_j$ where $i \neq j$, called Coriolis terms. The third type are the ones which involve only the first time derivative of the generalized coordinates and they are the dissipative forces due to the inherent damping. The fourth type of terms involve only p but not its derivatives. These arise from differentiating the potential energy. In the matrix-vector notation, the equations (2.77) are rewritten in a compact form as

$$M(p)\ddot{p} + C(p,\dot{p})\dot{p} + Dp + Kp = F. \tag{2.78}$$

The k,j-th element of the matrix $C(p,\dot{p})$ is defined as

$$\begin{aligned} c_{kj} &= \sum_{i=1}^{n} c_{ijk}(p)\dot{p}_i \\ &= \sum_{i=1}^{n} \frac{1}{2}\{\frac{\partial M_{kj}}{\partial p_i} + \frac{\partial M_{ki}}{\partial p_j} - \frac{\partial M_{ij}}{\partial p_k}\}\dot{p}_i. \end{aligned} \tag{2.79}$$

The systems represented by equation (2.78) satisfy an important property which is given in the next theorem.

Theorem (Property $\mathcal{S}$) The matrix $\mathcal{S}(p,\dot{p}) = \dot{M}(p) - 2C(p,\dot{p})$ is skew symmetric.

Proof The kj-th component of the time derivative of the inertia matrix, $\dot{M}(p)$ is given by the chain rule as

$$\dot{M}_{kj} = \sum_{i=1}^{n} \frac{\partial M_{kj}}{\partial p_i}\dot{p}_i.$$

Therefore, the kj-th component of $\mathcal{S} = \dot{M} - 2C$ is given by

$$\begin{aligned} \mathcal{S}_{kj} &= \dot{M}_{kj} - 2C_{kj} \\ &= \sum_{i=1}^{n} \left[\frac{\partial M_{kj}}{\partial p_i} - \left(\frac{\partial M_{kj}}{\partial p_i} + \frac{\partial M_{ki}}{\partial p_j} - \frac{\partial M_{ij}}{\partial p_k}\right)\right] \dot{p}_i \end{aligned}$$

$$= \sum_{i=1}^{n} \left[\frac{\partial M_{ij}}{\partial p_k} - \frac{\partial M_{ki}}{\partial p_j} \right] \dot{p}_i. \qquad (2.80)$$

Since the inertia matrix is symmetric, i.e., $M_{ij} = M_{ji}$, it follows from (2.80) by interchanging the indices k and j that

$$\mathcal{S}_{jk} = -\mathcal{S}_{kj}.$$

This completes the proof.

2.3.6 Summary and Remarks

A generic mathematical model for a class of multibody flexible spacecraft was developed. A judicious choice of coordinate systems was made which allowed effective decoupling of translational and rotational dynamics of the spacecraft. This is a useful way of modeling spacecraft dynamics since, depending on the application, one can choose to use only rotational, or translational, or a complete rotational-plus-translational dynamic model of the spacecraft. In the derivation of potential energy function, an assumption was made that the potential energy terms are only due to effects of elastic deformations; however, as stated previously, the potential energy contributions from any other sources of deformations, such as thermal deformations, can also be included in the same way as elastic deformations. The model developed can be used for multibody spacecraft such as space-based manipulators, multipayload platforms, and satellites with flexible appendages. With minor modifications, the model can be used even for terrestrial multibody systems such as multi-link robotic systems. In summary, the most spacecraft models and flexible manipulator models can be obtained as special cases of the model developed in this chapter.

Chapter 3

Dissipativity and Passivity

This chapter describes the concepts of dissipativity and passivity for linear as well as nonlinear dynamic systems and gives some fundamental stability results based on these notions. The topics covered in this chapter will form the background for the ensuing chapters.

3.1 Introduction

The notion of passivity was first introduced in the network theory literature. In the context of electrical networks, passivity has the implication that any single-port network consisting solely of resistors, capacitors, and inductors constitutes a passive system. Similary, for mechanical systems, any spring-mass-damper system with non-negative damping coefficients is passive. A more general form of passivity called "dissipativity" was introduced by Willems [Wil.72] and was investigated further in a number of articles (e.g., [Hil.76], [Hil.77], [Byr.91], and [Hil.92]). The notion of dissipativity is closely related to the rather intuitive phenomena of loss or dissipation of energy of a physical system. In short, any physical system, whose net energy gain remains non-positive at all times, is dissipative. Before we formalize these concepts, some mathematical notations and definitions are necessary and are given in Section 3.2. Section 3.3 gives various definitions of dissipative linear and nonlinear systems and their properties.

Section 3.4 presents some results on the stability of feedback interconnection of two passive systems.

3.2 Mathematical Preliminaries

In this section, we shall introduce the notations and build the mathematical background necessary for the development of the formal framework.

$\mathcal{L}_p$ and $\mathcal{L}_{pe}$ spaces:

Let $\mathcal{X}$ denote the set of all measurable real-valued n-vector functions $x(t)$ of time t, which map $\Re_+$ into $\Re^n$, where $\Re$ and $\Re_+$ denote the set of real numbers and the interval $[0,\infty)$, respectively, and $x(t) = (x_1(t)\ x_2(t)\ \cdots\ x_n(t))^T$. Suppose $\mathcal{L}_p^n = \mathcal{L}_p^n[0,\infty)$ denotes the set of all $x(t) \in \mathcal{X}$ such that

$$\int_0^\infty \mid x_i(t) \mid^p dt < \infty$$

where p is a positive integer. The set $\mathcal{L}_p^1$ ($:= \mathcal{L}_p$) forms a vector space over the field of real numbers $\Re$, and the p-norm of $x_i \in \mathcal{L}_p$ is defined as

$$\|x_i(t)\|_p = [\int_0^\infty \mid x_i(t) \mid^p)dt]^{\frac{1}{p}}$$

for finite p, and

$$\|x_i(t)\|_\infty = ess.\ \sup_{t\in\Re_+} \mid x_i(t) \mid .$$

The set $\mathcal{L}_p^n$ also forms a vector space over $\Re$, and the p-norm of $x(t)$ is defined as [Vid.93]:

$$\|x(t)\|_p = [\sum_{i=1}^n \|x_i(t)\|_p^2]^{\frac{1}{2}}.$$

Truncation:

Let $x(t) : \Re_+ \rightarrow \Re^n$. Then the truncation operator, P_T, is defined as the function

$$P_T(x(t)) = \begin{cases} x(t) & for\ t \leq T, \\ 0 & for\ t > T. \end{cases}$$

For simplicity, $P_T(x(t))$ is denoted by $x_T(t)$. For a fixed integer $p \in [1,\infty)$, the set $\mathcal{L}_{pe}^n = \mathcal{L}_{pe}^n[0,\infty)$ denotes the set of all functions $x(t)$ whose truncations

$x_T(t)$ are in $\mathcal{L}_p^n$ for all finite T. $\mathcal{L}_{pe}^n$ is called the extension of the space $\mathcal{L}_p^n$. The truncated p–norm of $x(t)$ is defined as: $\|x\|_{T_p} = \|x_T\|_p$.

For $p = 2$, the inner product of x and y in $\mathcal{L}_2^n$ is defined as:

$$< x, y >= \int_0^\infty x^T(t)y(t)dt.$$

The truncated inner product $< x, y >_T$ is defined as:

$$< x, y >_T = < x_T, y_T >= \int_0^T x^T(t)y(t)dt.$$

3.3 Dissipative Dynamic Systems

Consider a causal dynamic system Σ which maps input vectors $u(t) \in \mathcal{U}$ $(= \mathcal{L}_{pe}^m)$ into output vectors $y(t) \in \mathcal{Y}$ $(= \mathcal{L}_{pe}^l)$, i.e., $\Sigma : \mathcal{U} \to \mathcal{Y}$. Following [Hil.94], some definitions which will be used for further discussions of dissipativity-based properties of such systems, are given next.

Input-Output Dissipativity

Definition: The dynamic system Σ is said to be dissipative (in the input-output sense) with respect to the triplet (Q, S, R) where $Q = Q^T \in \Re^{l\times l}$, $R = R^T \in \Re^{m\times m}$ and $S \in \Re^{l\times m}$, if there exists a constant β such that

$$\int_0^T w(y,u)\,dt \quad := \quad < y, Qy >_T + 2 < y, Su >_T + < u, Ru >_T + \beta \geq 0$$
$$\forall T \geq 0, \;\; \forall u \in \mathcal{L}_{2e}^m.$$

The function $w(u, y)$ defined by the quadratic form

$$w(u,y) = [y^T \quad u^T] \begin{bmatrix} Q & S \\ S^T & R \end{bmatrix} \begin{bmatrix} y \\ u \end{bmatrix} \tag{3.1}$$

is called the "supply rate". In general, it is not necessary for $w(u, y)$ to be a quadratic function. For our purpose, however, we shall assume it to be quadratic.

Internal Dissipativity

Consider a nonlinear finite-dimensional system described by the n-dimensional state-space model:

$$\begin{aligned} \dot{x} &= f(x,u) \\ y &= h(x,u) \end{aligned} \tag{3.2}$$

where f, g are smooth (C^∞) functions and $f(0,0) = 0$; $h(0,0) = 0$. The internal dissipativity property for such systems is defined as follows.

Definition: A dynamic system is said to be internally dissipative with supply rate $w(u, y)$ if there exists a nonnegative function, $E : \mathcal{X} \rightarrow \Re$, called "storage function", such that $\forall\, T \geq 0$, $\forall x \in \mathcal{X}$, and $\forall u \in \mathcal{U}$,

$$E(x(0)) + \int_0^T w(u, y)dt \geq E(x(T)).$$

($\mathcal{X}$ denotes the set of $x(t)$ satisfying (3.2) $\forall\ u \in \mathcal{U}, \forall x(0) \in \Re^n$). The above inequality is called "dissipation inequality". The notion of storage function is a generalization of the concept of stored energy of the system. A more detailed treatment of these concepts and related results can be found in [Wil.72], [Hil.76], [Hil.92]. Equivalence of input-output (IO) dissipativity and internal dissipativity can be established if the system has additional properties of reachability and zero-state observability, which are defined next.

Definitions:

Let $\Phi(t, x_0, u)$ denote the state transition map for initial state x_0 at $t = 0$ and input $u(t)$. A function $\sigma(.)$ is said to belong to class K if it is strictly monotonic increasing and $\sigma(0) = 0$. Let $\mathcal{B}_r$ denote the set $\{x \in \Re^n, ||x|| < r\}$ where $||\cdot||$ denotes the Euclidian norm.

Reachability: A system Σ is reachable if there exist a constant $r > 0$ and a class K function σ such that, for every $x \in \mathcal{B}_r$, there exist a finite $t_1 \geq 0$ and an input u, $||u||_\infty \leq \sigma(||x||)$, such that $x = \Phi(t_1, 0, u)$. If this holds $\forall x \in \Re^n$, Σ is globally reachable.

We shall assume throughout that the system under consideration is globally reachable.

Zero-State Detectability - A system Σ is locally zero-state detectable if, with $u(t) \equiv 0$, there exists a neighborhood Ω of 0 such that $\forall x_0 \in \Omega$

$$y = h(x) = h(\Phi(t, x_0, 0)) = 0 \quad \forall t \geq 0 \Rightarrow \lim_{t\rightarrow\infty} x(t) = 0. \tag{3.3}$$

Also, if $\Omega = \Re^n$ system is said to be zero-state detectable (or globally zero-state detectable). A stronger property, namely zero-state observability, is defined next.

Zero-State Observability - A system, Σ is said to be locally zero-state observable if, with $u(t) \equiv 0$, there exists a neighborhood Ω of 0 such that $\forall x \in \Omega$

$$y = h(\Phi(t, x_0, 0)) = 0 \quad \forall t \geq 0 \Rightarrow x = 0. \tag{3.4}$$

Also, if $\Omega = \Re^n$ system is said to be zero-state observable (or globally zero-state observable).

It was proved in [Hil.92] that IO dissipativity is equivalent to internal dissipativity for zero-state observable systems.

The gain of a system is defined as follows [Vid.81].

Definition: The gain of the system (operator) $\Sigma : \mathcal{U} \rightarrow \mathcal{Y}$ is defined as:

$$\gamma_{p_B}(\Sigma) = \inf_{\substack{u \in \mathcal{U} \\ T \geq 0}} K \text{ such that } ||\Sigma u||_{T_p} \leq K||u||_{T_p} + B \text{ for some } \textit{finite } B.$$

B is called the bias, and γ_{p_B} is called the $\mathcal{L}_p$-gain with bias B.

$\mathcal{L}_p$-gain with zero bias, $\gamma_p(\Sigma)$, is defined as:

$$\gamma_p(\Sigma) = \inf_{\substack{u \in \mathcal{U} \\ T \geq 0}} K \text{ such that } ||\Sigma u||_{T_p} \leq K||u||_{T_p}.$$

The system Σ is said to be $\mathcal{L}_p$-stable if $u \in \mathcal{L}_p^m[0, \infty) \Rightarrow y \in \mathcal{L}_p^l[0, \infty)$. Σ is said to be finite-gain $\mathcal{L}_p$-stable if finite $\gamma_{p_B}(\Sigma)$ and B exist. Two commonly used definitions of stability are bounded-input-bounded-output (BIBO) stability which corresponds to $\mathcal{L}_\infty$-stability, and $\mathcal{L}_2$-stability, which implies that every finite-energy input produces a finite-energy output.

3.3.1 Passive Systems

Passive systems represent an important subset of dissipative systems. In particular, suppose $l = m$, $Q = 0$, $R = 0$, and $S = \frac{1}{2}I$ in Eq. (3.1). A system that is dissipative with respect to (w.r.t.) this supply rate satisfies: $< y, u >_T \geq 0 \; \forall T \geq 0$, $\forall u \in \mathcal{L}_{2e}^m$. This definition of passivity is in the input-output sense. As in the case of dissipativity, passivity can also be defined in the internal sense for finite-dimensional systems. Following [Hil.94], the formal definitions are given below. (We shall use the notation $||\cdot||$ and $||\cdot||_T$ to denote 2-norms unless stated otherwise. The notation $||\cdot||$ is sometimes used for Euclidian norm, but this usage should be clear from the context).

3.3.1.1 Input-Oputput Passivity

Passivity in the input-output sense can be defined for a large class of systems, including distributed-parameter systems.

Passivity: A system is said to be passive if it is dissipative w.r.t. the triplet $(0, \frac{1}{2}I, 0)$, i.e., if there exists a constant β such that

$$< u, y >_T + \beta \geq 0 \; \forall u \in \mathcal{L}_{2e}^m, \; \forall T \geq 0 \tag{3.5}$$

The following definitions represent different types of passivity, defined in the input-output sense [Hil.94].

General Passivity (GP): A system is said to be "general passive" if it is dissipative with respect to the triplet $(Q, S, R) = (-\delta I, \frac{1}{2}I, -\epsilon I)$ for some non-negative δ and ϵ, i.e., if $\exists$ a constant β and constants $\delta \geq 0$, $\epsilon \geq 0$ such that

$$< y, u >_T + \beta \geq \epsilon \|u\|_T^2 + \delta \|y\|_T^2 \quad \forall u \in \mathcal{L}_{2e}^m, \; \forall T \geq 0. \tag{3.6}$$

Input Strict Passivity (ISP): A system is said to be "input strictly passive" if it is dissipative with respect to the triplet $(Q, S, R) = (0, \frac{1}{2}I, -\epsilon I)$ for some $\epsilon > 0$, i.e., if $\exists$ a constant β and a constant $\epsilon > 0$ such that

$$< y, u >_T + \beta \geq \epsilon \|u\|_T^2 \;\; \forall u \in \mathcal{L}_{2e}^m, \; \forall T \geq 0. \tag{3.7}$$

Input strict passivity is often simply called **"strict passivity"**.

Output Strict Passivity (OSP): A system is said to be "output strictly passive" if it is dissipative with respect to the triplet $(Q, S, R) = (-\delta I, \frac{1}{2}I, 0)$ for some $\delta > 0$, i.e., if $\exists$ a constant β and a constant $\delta > 0$ such that

$$< y, u >_T + \beta \geq \delta \|y\|_T^2 \;\; \forall u \in \mathcal{L}_{2e}^m, \; \forall T \geq 0. \tag{3.8}$$

3.3.1.2 Internal Passivity

For the finite-dimensional case (i.e., systems described by Eq. (3.2)), passivity can be defined in the internal sense, as a special case of internal dissipativity. The difference between the input/output definitions and the internal definitions is that the latter additionally require the existence of a storage function which is a function of the state vector.

Internal Passivity: A system is said to be internally passive if there exists a nonnegative storage function $E[.]$ such that

$$< u, y >_T \geq E[x(T)] - E[x(0)] \;\; \forall u \in \mathcal{L}_{2e}^m, \; \forall T \geq 0. \tag{3.9}$$

General Passivity (internal): A system is said to be internally general passive if there exist non-negative numbers δ and ϵ and a non-negative storage function $E[.]$ such that

$$< y, u >_T \geq E[x(T)] - E[x(0)] + \epsilon ||u||_T^2 + \delta ||y||_T^2 \;\; \forall u \in \mathcal{L}_{2e}^m. \tag{3.10}$$

Input Strict Passivity (internal): A system is said to be internally input strictly passive if there exists an $\epsilon > 0$ and a non-negative storage function $E[.]$ such that

$$< y, u >_T \geq E[x(T)] - E[x(0)] + \epsilon ||u||_T^2 \;\; \forall u \in \mathcal{L}_{2e}^m, \; \forall T \geq 0. \tag{3.11}$$

Output Strict Passivity (internal): A system is said to be internally output strictly passive if there exists a $\delta > 0$ and a non-negative storage function $E[.]$ such that

$$< y, u >_T \geq E[x(T)] - E[x(0)] + \delta ||y||_T^2 \;\; \forall u \in \mathcal{L}_{2e}^m, \; \forall T \geq 0. \tag{3.12}$$

In addition to the above definitions, "state strict passivity" [Hil.94] is defined for such systems.

State Strict Passivity (SSP): A system is said to be "state strictly passive" if it is internally passive with storage function $E[.]$ and there exists a positive definite function $d(x)$, such that

$$< y, u >_T = E[x(T)] - E[x(0)] + \int_0^T d(x(t))dt \;\; \forall u \in \mathcal{L}_{2e}^m, \; \forall T \geq 0. \tag{3.13}$$

We shall refer to function $d(x)$ as the dissipation function since it represents the energy dissipation rate of the system.

3.3.2 Dissipative Nonlinear Systems Affine in Control

Consider a class of nonlinear systems affine in control, described by:

$$\begin{aligned} \dot{x} &= f(x) + g(x)u \\ y &= h(x) + N(x)u \end{aligned} \tag{3.14}$$

where $x \in \Re^n$, $u \in \mathcal{U}$, $y \in \mathcal{Y}$, $f(x) \in \Re^n$, $h(x) \in \Re^l$, $g(x) \in \Re^{n \times m}$, $N(x) \in \Re^{l \times m}$, with f, g, h, N smooth, $f(0) = 0$, and $h(0) = 0$. Suppose the supply rate is given by

$$w(u, y) = y^T Q y + 2y^T S u + u^T R u. \tag{3.15}$$

It is assumed that the system is globally reachable. The following thorem [Hil.76] gives the necessary and sufficient conditions for system (3.14) to be internally dissipative.

Theorem 3.1 *The nonlinear system (3.14) is internally dissipative with respect to supply rate w, given by Eq. (3.15), if and only if there exist functions $E(x) \in C^1$, $l(x) \in \Re^k$, and $W(x) \in \Re^{k \times m}$ for some integer k such that*

$$\begin{aligned} E(x) \geq 0, & \qquad E(0) = 0 \\ \nabla^T E(x) f(x) &= h^T(x) Q h(x) - l(x)^T l(x) \\ \frac{1}{2} g^T(x) \nabla E(x) &= \overline{S}^T(x) h(x) - W^T(x) l(x) \\ \overline{R}(x) &= W^T(x) W(x) \\ \overline{S}(x) &= Q N(x) + S \quad \textit{and} \\ \overline{R}(x) &= R + N^T(x) S + S^T N(x) + N^T(x) Q N(x) \end{aligned} \tag{3.16}$$

where $\nabla E(x)$ denotes $\frac{\partial E(x)}{\partial x}$.

Proof 3.1 The proof of this theorem can be found in [Hil.76]. □

The following result can be obtained from Theorem 3.1.

Corollary: *If system (3.14) is dissipative with supply rate $w(u, y)$ in (3.15) then there exists a storage function $E(x)$ such that $E(0) = 0$, $E(x) \geq 0$, and*

$\forall u \in \mathcal{L}_{2e}^m$,

$$\begin{aligned} \frac{dE(x)}{dt} &= -d(x,u) + w(u,y), \qquad (3.17) \\ \textit{where } d(x,u) &= [l(x) + W(x)u]^T [l(x) + W(x)u]. \end{aligned}$$

Theorem 3.1 gives conditions for internal dissipativity of nonlinear systems. Some special cases of interest are given next.

Special Cases:

I) **Passivity:** If the system (3.14) is dissipative with respect to quadratic supply rate $w(u,y)$, given by (3.1), where $Q = 0$, $R = 0$ and $S = \frac{1}{2}I$, the conditions of Theorem 3.1 reduce to the nonlinear version of the well-known Kalman-Yakubovich lemma [Hil.92, Hil.94, Byr.91].

Theorem 3.2 *The nonlinear system (3.14) is internally passive if and only if there exists a non-negative function $E(x) \in C^1$, $E(0) = 0$, and functions $l(x) \in \Re^k$, and $W(x) \in \Re^{k\times m}$ for some integer k such that*

$$\begin{aligned} \nabla^T E(x) f(x) &= -l^T(x) l(x) \\ g^T(x) \nabla E(x) &= h(x) - 2W^T(x) l(x) \\ W^T(x) W(x) &= \frac{1}{2}[N(x) + N^T(x)]. \qquad (3.18) \end{aligned}$$

Proof 3.2 The proof of this theorem is obtained by straightforward application of Theorem 3.1. □

Note that the Eqs. (3.18) are analogous to the Kalman-Yakubovich lemma in the linear time-invariant case. For internally passive systems, if $N(x) = 0$, then $W(x) = 0$ and $d(x,u) := d(x) = -l^T(x)l(x)$ for that case. Equations (3.17) can be equivalently stated as

$$E[x(T)] - E[x(0)] = \int_0^T y^T(t)u(t)dt - \int_0^T d(x(t))dt \quad \forall u \in \mathcal{L}_{2e}^m, \forall T \geq 0. \qquad (3.19)$$

There are two sub-cases of passivity depending on the sign definiteness of $d(x)$. i) If $d(x) > 0$ (positive definite) system (3.14) is *state strictly passive* and it satisfies (3.13).

ii) If $d(x) = 0$ system (3.14) is called *lossless passive*, and it satisfies

$$E[x(T)] = E[x(0)] + \int_0^T y^T(t)u(t)dt. \tag{3.20}$$

II) **Finite Gain Systems:** If system $\Sigma : \mathcal{U} \to \mathcal{Y}$ is dissipative with respect to the supply rate (3.1) with $Q = -I$, $R = \mu^2 I$, $S = 0$, we have

$$||y||_T \leq \mu ||u||_T.$$

The above inequality is satisfied for $\mu = \gamma(\Sigma)$, the $\mathcal{L}_2$-gain with zero bias. That is, the system Σ is dissipative with respect to $(Q, S, R) = (-I, 0, \gamma^2(\Sigma)I)$. From Theorem 3.1, the system is internally dissipative with respect to this supply function if and only if the following equations are satisfied for some non-negative $E(x) \in C^1$ with $E(0) = 0$, and $l(x) \in \Re^k$, $W(x) \in \Re^{k\times m}$ for some integer k.

$$\begin{aligned} \nabla^T E(x) f(x) &= -h^T(x)h(x) - l^T(x)l(x) \\ \frac{1}{2} g^T(x)\nabla E(x) &= -N^T(x)h(x) - W^T(x)l(x) \\ \mu^2 I - N^T(x)N(x) &= W^T(x)W(x). \end{aligned} \tag{3.21}$$

The above conditions can be derived from Theorem 3.1, and represent the nonlinear version of the Bounded-Real Lemma [Hil.94].

In the case of linear time-invariant (LTI) systems, dissipativity has both, frequency- and time-domain, interpretations. In the next section, we will give important definitions and properties of finite dimensional linear time-invariant (FDLTI) dissipative systems.

3.3.3 Passive FDLTI Systems

For finite-dimensional linear, time-invariant (FDLTI) systems, input-output passivity is equivalent to internal passivity of a minimal realization. The equivalence is a result of the Kalman-Yakubovich lemma. Therefore, in the discussion of FDLTI systems, we shall use the term *passivity* to represent input/output passivity. For such systems, passivity is equivalent to "positive realness" of the transfer function [New.66, Des.75]. The concept of strict positive realness has also been defined in the literature, and is closely related to strict passivity. An important application of *strict* passivity is in the investigation of stability of feedback interconnection of two passive systems. In particular, the Passivity Theorem [Des.75] states that the negative feedback interconnection of a passive system and a strictly passive (i.e., input-strictly passive) finite-gain system is $\mathcal{L}_2$-stable. The passivity theorem can be directly applied to FDLTI systems to show that the feedback interconnection of a positive real (PR) system and a stable strictly passive FDLTI system is stable. Although positive realness of FDLTI systems is equivalent to passivity, the relationship between *strict* passivity and *strict* positive realness is somewhat more complicated because there are several definitions of *strict* positive realness. Strict positive realness (SPR) has been defined in the literature only for stable FDLTI systems (with all poles in the open left-half plane). Strict passivity is equivalent to the strongest definition of SPR [Tay.74] for such systems. However, the requirement of strict passivity is too stringent, as it includes only systems with a relative degree of zero and thus excludes a large classs of systems. Another definition of strict positive realness was introduced in [Tay.74] for scalar systems, and was further investigated in [Ioa.87, Tao.88, Wen.88a] for multivariable systems. This definition (referred to as "strong SPR") is weaker than strict passivity. An even less stringent definition of SPR, termed "weak SPR", was introduced in [Loz.90], which allows a larger class of strictly proper systems. Recently, the definition of marginally strictly positive real (MSPR) was presented in [Jos.94, Jos.96], which allows the system to have poles *on* the imaginary axis, and is therefore the weakest definition of SPR systems. This section presents various definitions of SPR systems, a minimal realization of MSPR systems, and an extension of the Kalman-Yakubovich lemma to such systems, and sets the stage for the stability results presented in the subsequent section.

Let $G(s)$ denote an $m \times m$ matrix whose elements are proper rational functions of the complex variable s. The matrix $G(s)$ is said to be stable if every element $G_{ij}(s)$ is analytic in $Re(s) \geq 0$. Let the conjugate-transpose of a complex matrix T be denoted by T^*.

Definition 1: An $m \times m$ rational matrix $G(s)$ is said to be *positive real* (PR) if

(i) all elements of $G(s)$ are analytic in $Re(s) > 0$;

(ii) $G(s) + G^*(s) \geq 0$ in $Re(s) > 0$; or equivalently,

(iia) poles on the imaginary axis are simple and their residues are non-negative definite, and

(iib) $G(j\omega) + G^*(j\omega) \geq 0$ for $\omega \in (-\infty, \infty)$.

Some definitions of SPR systems that are found in the literature (e.g., [Tay.74, Ioa.87, Tao.88, Wen.88a, Loz.90]) are given next. Definition 2, which represents the specialization to LTI systems, of the general definition of strict passivity, is the strongest definition of strict positive realness.

Definition 2: An $m \times m$ stable rational matrix $G(s)$ is said to be *strictly passive* if there exists an $\epsilon > 0$ such that

$G(j\omega) + G^*(j\omega) \geq \epsilon I$ for $\omega \in (-\infty, \infty)$.

Definition 3: An $m \times m$ stable rational matrix $G(s)$ is said to be *strictly positive real in the strong sense* (strong SPR, or SSPR) if $G(s - \epsilon)$ is PR for some $\epsilon > 0$; that is, if

(i) $G(j\omega) + G^*(j\omega) > 0$ for $\omega \in (-\infty, \infty)$

(ii) $\mathcal{Z} = G(\infty) + G^T(\infty) \geq 0$

(iii) $\lim_{\omega \to \infty} \omega^2 [G(j\omega) + G^*(j\omega)] > 0$ if $\mathcal{Z}$ is singular.

Definition 4: An $m \times m$ stable rational matrix $G(s)$ is said to be *strictly positive real in the weak sense* (weak SPR, or WSPR) if

$$G(j\omega) + G^*(j\omega) > 0 \; for \; \omega \in (-\infty, \infty).$$

Note that Definition 2 requires that $\mathcal{Z} = G(\infty) + G^T(\infty)$ to be positive definite; i.e., the system must have a relative degree of zero. This requirement makes the definition of strictly passive systems too restrictive. Definition 3 (SSPR) can include certain strictly proper systems which satisfy additional conditions (ii) and (iii). Definition 4 (WSPR) does not require these additional conditions, and is therefore less restrictive than Definition 3. However, all the definitions 2-4 assume $G(s)$ to be stable. The following definition of SPR, introduced in [Jos.94, Jos.96], allows $G(s)$ to have poles *on* the imaginary axis.

Definition 5: An $m \times m$ rational matrix $G(s)$ is said to be *marginally strictly positive real* (MSPR) if it is positive real, and

$$G(j\omega) + G^*(j\omega) > 0 \; for \; \omega \in (-\infty, \infty).$$

Definition 5 of MSPR differs from Definition 1 (PR) because the frequency domain inequality ($\geq$) has been replaced by the strict inequality ($>$). The difference between Definitions 4 and 5 is that the latter allows $G(s)$ to have poles on the imaginary axis. This is an important difference because many real-life systems contain pure integrators and oscillators, which are permitted under Definition 5, but not under Definitions 2, 3, and 4. Suppose $[A, B, C, D]$ is a minimal realization of a proper rational matrix $M(s)$. $M(s)$ (or $[A, B, C, D]$) is said to be minimum-phase if its transmission zeros are confined to the open left-half plane (OLHP); i.e., the rank of $\begin{bmatrix} sI - A & B \\ -C & D \end{bmatrix}$ can drop below its normal value only for values of s in the OLHP.

Next, we shall state some important properties of MSPR systems since they will be used in the stability proofs later when we address the stability of feedback interconnections.

3.3.3.1 Properties of MSPR Systems

Suppose $G(s)$ is positive real and has all poles (i.e., the eigenvalues of the system matrix of its minimal realization) in the closed left-half plane. Following [And.67], $G(s)$ can be written as:

$$G(s) = G_1(s) + G_2(s). \tag{3.22}$$

where $G_1(s)$ has purely imaginary poles, and $G_2(s)$ has poles only in the OLHP. Furthermore, $G_1(s)$ is of the form:

$$G_1(s) = \frac{\alpha_0}{s} + \sum_{i=1}^{p} \frac{\alpha_i s + \beta_i}{s^2 + \omega_i^2}. \tag{3.23}$$

where α_i and β_i are $m \times m$ real matrices, and $\omega_i > 0, i = 1, 2, ..., p$ ($\omega_i \neq \omega_j$ for $i \neq j$).

Some remarks regarding the nature of the poles on the imaginary axis are in order. The poles and zeros considered here are in the Smith-McMillan sense [Kai.80]; i.e., there can be more than one pole at a given location, without it being considered a "repeated" pole. In particular, using standard results in matrix fraction descriptions [Kai.80], it can be shown that the McMillan degree (i.e., the minimal order of a state space representation) of the term: $[\alpha_0/s]$ is equal to $\rho(\alpha_0)$, where $\rho(.)$ denotes the rank. That is, there are $\nu = \rho(\alpha_0)$ simple poles at $s = 0$.

Suppose the McMillan degree of the term: $[\alpha_i s + \beta_i]/(s^2 + \omega_i^2)$ is $2k_i$, where $k_i \leq m$. Then this term has k_i simple poles (each) at $s = j\omega_i$ and $s = -j\omega_i$.

The following result states that $G(s)$ is PR (respectively MSPR) iff its stable part is PR (resp. WSPR).

Lemma 1 *$G(s)$ is PR (resp. MSPR) iff all of the following hold:*

(i) $G_2(s)$ is PR (resp. WSPR)

(ii) $\alpha_i = \alpha_i^T \geq 0, i = 0, 1, 2, ...p$

(iii) $\beta_i = -\beta_i^T, i = 1, 2, ...p$.

Proof 1 From the requirement that the residues at the imaginary-axis poles be nonnegative-definite, we get (ii) and (iii) (See [New.66]). Therefore, we have: $G_1(j\omega) + G_1^*(j\omega) = 0$, and $G(j\omega) + G^*(j\omega) = G_2(jw) + G_2^*(j\omega)$, which is positive semi-definite (resp. positive definite) for all real ω iff $G_2(s)$ is PR (resp. WSPR). □

If $G(s)$ is MSPR, the degree of $[G(s) + G^*(s)]$ is generally less than $2\times$ degree of $G(s)$; also, $G_2(s)$ is stable, and both $G(s)$ and $G_2(s)$ are minimum-phase. We shall next consider a minimal realization of MSPR transfer functions.

3.3.3.2 Minimal Realization of MSPR Systems

Consider the realization of $G_1(s)$. For the term $[\alpha_0/s]$, since α_0 is symmetric and non-negative definite (Lemma 1), there exists an $m \times m$ real orthogonal transformation matrix T which diagonalizes it, i.e.,

$$T^T \alpha_0 T = \text{diag}[\lambda_1, \lambda_2, \ldots, \lambda_\nu, 0, \ldots, 0] \tag{3.24}$$

where λ_i are positive scalars. Let Λ denote $diag[\lambda_1, \lambda_2, \ldots, \lambda_\nu]$. A minimal realization (of order ν) of $T^T[\alpha_0/s]T$ is: $[\mathcal{A}_0, \overline{\mathcal{B}}_0, \overline{\mathcal{C}}_0, 0]$, where

$$\mathcal{A}_0 = [0_\nu];\ \overline{\mathcal{B}}_0 = [I_\nu\ 0];\ \overline{\mathcal{C}}_0 = \begin{bmatrix} \Lambda \\ 0 \end{bmatrix} \tag{3.25}$$

where 0_ν and I_ν denote $\nu \times \nu$ null and identity matrices. Therefore, a minimal (ν^{th}-order) realization of $[\alpha_0/s]$ is given by $[\mathcal{A}_0, \mathcal{B}_0, \mathcal{C}_0, 0]$, where

$$\mathcal{B}_0 = \overline{\mathcal{B}}_0 T^T; \mathcal{C}_0 = T\overline{\mathcal{C}}_0 \tag{3.26}$$

Considering the term: $[\alpha_i s + \beta_i]/(s^2 + \omega_i^2)$, if its McMillan degree is $2k_i(k_i \leq m)$, a minimal realization is given by [And.67]: $[\mathcal{A}_i, \mathcal{B}_i, \mathcal{B}_i^T, 0]$, where $\mathcal{A}_i \in \Re^{2k_i \times 2k_i}$,

$$\mathcal{A}_i = \oplus_{j=1}^{k_i} \Xi_j \tag{3.27}$$

where $\oplus$ denotes the direct sum, and

$$\Xi_j = \begin{bmatrix} 0 & -\omega_j \\ \omega_j & 0 \end{bmatrix} \tag{3.28}$$

and $\mathcal{B}_i \in \Re^{2k_i \times m}$. Then a minimal realization of $G_1(s)$ is given by: $[A_1, B_1, C_1, 0]$, where $A_1 = \mathcal{A}_0 \oplus \mathcal{A}_1 \oplus \ldots \mathcal{A}_p$; $B_1 = [\mathcal{B}_0^T, \mathcal{B}_1^T, \ldots, \mathcal{B}_p^T]^T$; $C_1 = [\mathcal{C}_0, \mathcal{B}_1^T, \ldots, \mathcal{B}_p^T]$; $A_1 \in \Re^{n_1 \times n_1}, B_1 \in \Re^{n_1 \times m}, C_1 \in \Re^{m \times n_1} (n_1 = \nu + 2\sum_{i=1}^{p} k_i)$

Let $[A_2, B_2, C_2, D]$ be a minimal realization $G_2(s)$, where $A_2 \in \Re^{n_2 \times n_2}$, $B_2 \in \Re^{n_2 \times m}, C_2 \in \Re^{m \times n_2}$, and $D \in \Re^{m \times m}$ $(n_2 = n - n_1)$. Let $A = A_1 \oplus A_2$; $B = [B_1^T, B_2^T]^T$; $C = [C_1, C_2]$. Then $A \in \Re^{n \times n}, B \in \Re^{n \times m}, C \in \Re^{m \times n}$, and $D \in \Re^{m \times m}$. Since $G_1(s)$ and $G_2(s)$ have no poles in common, $[A, B, C, D]$ is a minimal realization of $G(s)$.

3.3.3.3 Time-Domain Characterization of MSPR Systems

We next present the time-domain (state-space) characterization of MSPR systems, which is an extension of the Kalman-Yakubovich lemma for WSPR systems, that was proved in [Loz.90]. (Only the sufficient condition is given since it is of relevance in obtaining the stability results presented in Section 3.4).

Lemma 2 *If $G(s)$ is MSPR, there exist real matrices: $P = P^T > 0, P \in \Re^{n \times n}, \mathcal{L} \in \Re^{m \times n_2}, W \in \Re^{m \times m}$ such that*

$$A^T P + PA = -L^T L \tag{3.29}$$

$$C = B^T P + W^T L \tag{3.30}$$

$$W^T W = D + D^T \tag{3.31}$$

$$L = [0_{m \times n_1}, \mathcal{L}_{m \times n2}] \tag{3.32}$$

where $[A_2, B_2, \mathcal{L}, W]$ is minimal and minimum-phase, i.e., $F(s) = W + L(sI - A)^{-1}B = W + \mathcal{L}(sI - A_2)^{-1}B_2$ is minimum-phase.

Proof 2 $G(s)$ has the form given by Eqs. (3.22) and (3.23). Since $G(s)$ is MSPR, conditions (i)-(iii) of Lemma 1 hold. Considering $G_1(s)$ in Eq. (3.23), it consists of $(p+1)$ transfer functions in parallel. A minimal realization (Eqs. 3.25 and 3.26) of the first term, $[\alpha_0/s]$, is: $[\mathcal{A}_0, \mathcal{B}_0, \mathcal{C}_0, 0]$. Letting $\Pi_0 = \Lambda$, it can be verified that the following equations are satisfied:

$$\mathcal{A}_0^T \Pi_0 + \Pi_0 \mathcal{A}_0 = 0 \tag{3.33}$$

$$\Pi_0 \mathcal{B}_0 = \mathcal{C}_0^T. \tag{3.34}$$

A minimal realization (Eqs. 3.27, 3.28) of the i-th component of the second term, $[\alpha_i s + \beta_i]/(s^2 + \omega_i^2)$, is given by $[\mathcal{A}_i, \mathcal{B}_i, \mathcal{B}_i^T, 0]$. Letting $\Pi_i = I_{2k_i}$, it can be seen that this realization satisfies:

$$\mathcal{A}_i^T \Pi_i + \Pi_i \mathcal{A}_i = 0 \tag{3.35}$$

$$\Pi_i \mathcal{B}_i = \mathcal{B}_i. \tag{3.36}$$

Finally, following [Loz.90], since $G_2(s)$ is WSPR, there exist $P_2 = P_2^T > 0$, $P_2 \in \Re^{n_2 \times n_2}$, $W \in \Re^{m \times m}$, $\mathcal{L} \in \Re^{m \times n_2}$ such that

$$A_2^T P_2 + P_2 A_2 = -\mathcal{L}^T \mathcal{L} \tag{3.37}$$

$$C_2 = B_2^T P_2 + W^T \mathcal{L} \tag{3.38}$$

$$W^T W = D + D^T \tag{3.39}$$

where $[A_2, B_2, \mathcal{L}, \mathcal{W}]$ is minimal and minimum-phase. Defining

$$P_1 = \Pi_0 \oplus \Pi_1 \cdots \oplus \Pi_p \tag{3.40}$$

$$P = P_1 \oplus P_2. \tag{3.41}$$

we have the required result. □

3.4 Stability of Passive Systems

While addressing the issue of stability we need to make distinction between input-output (IO) stability and Lyapunov stability of the free system. IO stability can be defined as $\mathcal{L}_p$-stability or finite-gain stability. Finite-gain stability implies $\mathcal{L}_p$-stability. The dissipativity approach naturally leads to $\mathcal{L}_2$-stability; therefore we shall henceforth use the term "IO-stability" to imply $\mathcal{L}_2$-stability. The most useful form of Lyapunov stability is asymptotic stability. In general, for nonlinear systems, IO stability does not imply asymptotic stability or vice versa. However, in case of finite dimensional linear time-invariant (FDLTI) systems, the IO stability is equivalent to asymptotic stability of any minimal realization of the system. For FDLTI systems, passivity implies the existence of Lyapunov function (which is simply the storage function) $E(x) = x^T P x$, $P > 0$, such that $\frac{dE}{dt} \leq y^T u$, i.e., the corresponding free system ($u = 0$) is Lyapunov stable. When the inequality is strict the free system is asymptotically stable. Similarly, for nonlinear systems, from Eq. (3.17), if $d(x, 0) := d(x)$ is positive semidefinite and $E(x)$ is positive definite the free system is Lyapunov stable. If both $E(x)$ and $d(x)$ are positive definite, the free system is asymptotically stable.

It was proved in [Hil.76] that, if Σ is dissipative with respect to $w(u, y)$ and is zero-state-observable, then the storage function $E(x)$ is positive definite.

Consider a nonlinear system which is affine in control, given by (3.14) and the supply rate given by (3.1). We assume that the system is globally reachable. The following theorem gives the stability conditions for such system. (Throughout this book, we shall equivalently use the terms "stability of the origin" and "stability of the system".)

Theorem 3.3 *If the nonlinear system (3.14) is internally dissipative with respect to supply rate $w(u, y)$ in (3.1) then, under the assumptions of global reachability and zero-state observability, the corresponding free system, Σ_f given by $\dot{x} = f(x)$, is Lyapunov stable if $Q \leq 0$ and globally asymptotically stable if $Q < 0$.*

Proof 3.3 The proof can be found in [Hil.76]. □

Remarks - It can be shown [Hil.94] that passive systems and ISP systems are Lyapunov stable whereas OSP systems are asymptotically stable.

3.4.1 Stability of Feedback-Interconnected Systems

We shall now address the issue of stability of the feedback loop when two passive systems are connected in the feedback configuration. We shall address two different scenarios: one in which a passive linear system is controlled by a linear controller and the second in which a passive nonlinear system is controlled by a nonlinear controller as well as a linear controller. We shall present the stability theorems which, to the best of our knowledge, give the largest class of passive FDLTI controllers for stabilizing passive FDLTI systems and passive nonlinear systems. The next section addresses the stability of feedback interconnection of two FDLTI systems.

3.4.1.1 Feedback Interconnection of Passive LTI Systems

Consider the system in Figure 3.1, where $G(s)$ and $H(s)$ are $m \times m$ proper rational matrices. This composite system is always well-posed. The system is said to be asymptotically stable if its state-space realization consisting of individual minimal realizations of $G(s)$ and $H(s)$, is asymptotically stable. We have the following stability result [Jos.94, Jos.96].

Theorem 3.4 *The negative feedback interconnection of $G(s)$ and $H(s)$ is asymptotically stable if all of the following conditions are satisfied:*

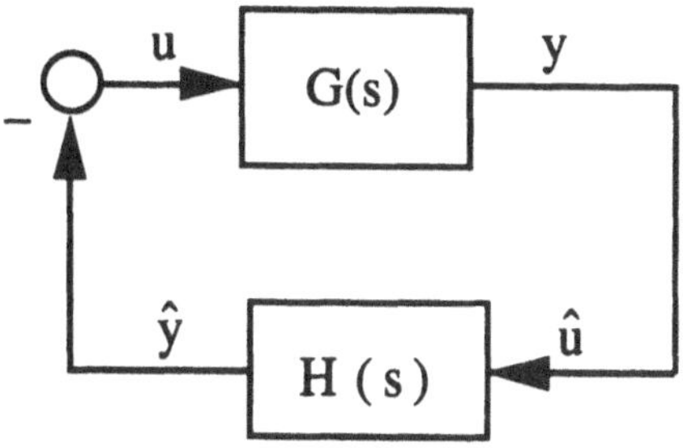

Figure 3.1: Negative feedback loop

(i) $G(s)$ is MSPR

(ii) $H(s)$ is PR

(iii) None of the $j\omega$ -axis poles of $G(s)$ is a transmission zero of $H(s)$

Proof 3.4 Let $[A, B, C, D]$ denote the minimal realization of $G(s)$ described in Section 3.3.3, and let x be the corresponding state vector of order n. Let n_2 denote the number of poles of the stable part $G_2(s)$ of $G(s)$, and let $[A_2, B_2, C_2, D]$ denote its minimal realization. Let $[\hat{A}, \hat{B}, \hat{C}, \hat{D}]$ denote a minimal realization of $H(s)$, and let $\hat{x}$ denote the corresponding $\hat{n}^{th}$ order state vector. Since $G(s)$ is MSPR, from Lemma 2, there exist matrices $P = P^T > 0, P \in \Re^{n \times n}, \mathcal{L} \in \Re^{m \times n_2}, W \in \Re^{m \times m}$, such that Eqs. (3.29)-(3.32) are satisfied, and $[A_2, B_2, \mathcal{L}, W]$ is minimal and minimum-phase. Since $H(s)$ is PR, there exist matrices $\hat{P} = \hat{P}^T > 0, \hat{P} \in \Re^{\hat{n} \times \hat{n}}, \hat{L} \in \Re^{\hat{k} \times \hat{n}}, \hat{W} \in \Re^{\hat{k} \times m}$, such that [And.67]

$$\hat{A}^T \hat{P} + \hat{P} \hat{A} = -\hat{L}^T \hat{L} \tag{3.42}$$

$$\hat{C} = \hat{B}^T \hat{P} + \hat{W}^T \hat{L} \tag{3.43}$$

$$\hat{W}^T \hat{W} = \hat{D} + \hat{D}^T. \tag{3.44}$$

Consider the candidate Lyapunov function

$$V(x, \hat{x}) = x^T P x + \hat{x}^T \hat{P} \hat{x}. \tag{3.45}$$

Proceeding as in the proof of Theorem 1 in [Loz.90] we have:

$$\dot{V} = 2u^T y - z^T z + 2\hat{u}^T \hat{y} - \hat{z}^T \hat{z} \tag{3.46}$$

where

$$z = Lx + Wu = \mathcal{L}x_2 + Wu \tag{3.47}$$

$$\hat{z} = \hat{L}\hat{x} + \hat{W}\hat{u}. \tag{3.48}$$

Since $\hat{u} = y$ and $\hat{y} = -u$,

$$\dot{V} = -z^T z - \hat{z}^T \hat{z} \leq 0 \tag{3.49}$$

i.e., $\dot{V}$ is negative semi-definite. Therefore, the closed-loop system is at least Lyapunov-stable. By examining the trajectories for which $\dot{V}(t) = 0$, it will now be shown that the system is asymptotically stable. The system is Lyapunov stable, therefore $u(t)$ and $y(t)$ can consist only of exponentially decaying terms, sinusoids, and constant terms. Now, $\dot{V}(t) = 0$ implies $z(t) = 0$; however, $z(t)$ is the output of the system: $F(s) = W + \mathcal{L}(sI - A_2)^{-1}B_2$, (produced by the input $u(t)$). $F(s)$ does not have zeros on the imaginary axis, therefore, $u(t)$ can contain only exponentially decaying terms. Therefore, $y(t)$ can contain only exponentially decaying terms and sinusoids at frequencies corresponding to the imaginary-axis poles of $G(s)$. Because of (*iii*), this would imply that $\hat{y}(t)$ [$= -u(t)$] can contain sinusoidal terms, which contradicts the fact that $u(t) \to 0$. Therefore, $y(t)$ must decay exponentially. Because of the minimality of $[A, B, C, D]$ and $[\hat{A}, \hat{B}, \hat{C}, \hat{D}]$, this implies $x(t) \to 0$, $\hat{x}(t) \to 0$ exponentially; i.e., $||\tilde{x}(t)|| \leq ||\tilde{x}(0)||c_1 e^{-c_2 t}$ for some positive c_1 and c_2, where $\tilde{x} = (x^T, \hat{x}^T)^T$. [An alternate proof that $\tilde{x} \to 0$ can be obtained by using (iii) and some results in polynomial matrix descriptions [Kai.80] to show that the $(n+\hat{n})$-th order closed-loop system is detectable with respect to output $z(t)$, and therefore, $z(t) = 0$ $\Rightarrow \tilde{x}(t) \to 0$]. Let λ_M and λ_m denote the largest and smallest eigenvalues of $[diag(P, \hat{P})]$. It can be shown that

$$V(t) \leq c_1^2[\lambda_M/\lambda_m]V(0)e^{-2c_2 t}$$

along all trajectories for which $\dot{V} = 0$. Therefore, $V(t)$ must *decrease* with t, and $\dot{V}$ must be negative for some t. That is, such trajectories cannot exist, and by LaSalle's invariance principle, the system is asymptotically stable. □

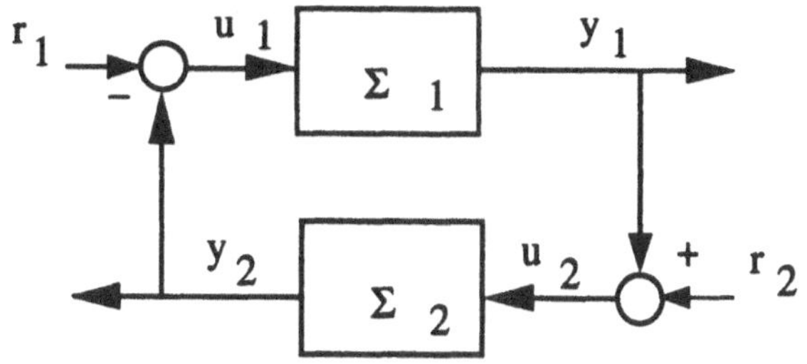

Figure 3.2: Feedback interconnection of nonlinear systems

It should be noted that $G(s)$ and $H(s)$ in Theorem 3.4 are completely interchangeable. The following corollaries follow from Theorem 3.4.

Corollary *The negative feedback interconnection of $G(s)$ and $H(s)$ is stable if $G(s)$ is WSPR and $H(s)$ is PR.*

The above corollary is the stability result given in [Loz.90] applied to the FDLTI case.

Corollary *The negative feedback interconnection of $G(s)$ and $H(s)$ is stable if both $G(s)$ and $H(s)$ are MSPR.*

The stability result of Theorem 3.4 states that any PR system can be stabilized by an MSPR controller. The significance of the results is that the stability does not depend on the plant order or the parametric values of the plant, and is therefore *robust*. In the next section, we shall consider the case of feedback-interconnected nonlinear passive systems.

3.4.1.2 Feedback Interconnection of Passive Nonlinear Systems

Consider the negative feedback interconnection of two nonlinear systems, Σ_1 and Σ_2, as shown in Figure 3.2. It is assumed that the composite system is well-posed. The IO stability of this system is defined as that of the map from (r_1, r_2) to (y_1, y_2). The following theorem, known as the "Passivity Theorem", gives a sufficient condition for finite-gain stability [Vid.93].

Theorem 3.5 *Consider feedback system of Figure 3.2. Suppose both* Σ_1 *and* Σ_2 *are general passive i.e., they satisfy*

$$\int_0^T y_i^T u_i dt \geq \beta_i + \epsilon_i \int_0^T u_i^T u_i dt + \delta_i \int_0^T y_i^T y_i dt \quad i = 1,2, \ \forall u \in \mathcal{U}, \ \forall T \geq 0 \tag{3.50}$$

for some β_i, $\epsilon_i \geq 0$, $\delta_i \geq 0$, $i = 1, 2$. *Then the feedback system is finite-gain stable if*

$$\epsilon_1 + \delta_2 > 0 \quad \text{and} \quad \epsilon_2 + \delta_1 > 0. \tag{3.51}$$

Corollary [Hil.94]: *The feedback system of Figure 3.2 is finite-gain stable if*

$$\Sigma_1, \ \Sigma_2 \ \ \text{are} \ \ ISP \ \ \text{or}$$

$$\Sigma_1, \ \Sigma_2 \ \ \text{are} \ \ OSP.$$

Another version of the Passivity Theorem, which addresses asymptotic stability, was proved in [Hil.94], and is given below. (Σ_1 and Σ_2 are assumed to be affine in control and have the usual smoothness properties. x_1 and x_2 denote their respective state vectors).

Theorem 3.6 *Suppose* Σ_1 *and* Σ_2 *in Figure 3.2 are zero-state observable and internally general passive, i.e., there exist positive definite storage functions,* $E_i(x_i)$ $(i = 1, 2)$ *such that*

$$\begin{aligned} \int_0^T y_i^T u_i dt \ \ &\geq \ \ E_i(x_i(T)) - E_i(x_i(0)) + \epsilon_i \int_0^T u_i^T u_i dt + \delta_i \int_0^T y_i^T y_i dt \\ &\forall u \in \mathcal{U}, \ \ \forall T \geq 0 \end{aligned} \tag{3.52}$$

along trajectories of the system. Then the origin $(0,0)$ *is asymptotically stable if* Σ_1 *and* Σ_2 *are zero-state observable and*

$$\epsilon_1 + \delta_2 > 0 \quad \text{and} \quad \epsilon_2 + \delta_1 > 0. \tag{3.53}$$

Proof 3.6 Refer to [Hil.94]. □

In addition to the conditions of the above theorem, if the storage functions $E_i(x_i)$ are radially unbounded, the origin of the system is globally asymptotically stable.

Corollary [Hil.94]: *If Σ_1 and Σ_2 are passive, and satisfy any one of the following conditions, then the origin of the closed-loop system is asymptotically stable.*

i) both Σ_1 and Σ_2 are ISP and zero-state observable

ii) both Σ_1 and Σ_2 are OSP and zero-state observable

iii) Σ_1 or Σ_2 is zero-state observable and either Σ_1 is OSP or Σ_2 is ISP.

iv) $(\Sigma_2\Sigma_1)$ is zero-state observable and either Σ_1 is ISP or Σ_2 is OSP.

The above theorems require one of the systems to be ISP or OSP, which is a rather stringent requirement. In order to weaken the strict passivity requirement, we shall next consider the stability of the feedback interconnection of a nonlinear passive system and a linear PR controller.

3.4.1.3 Feedback Interconnection of Passive Nonlinear/LTI Systems

Consider the system shown in Figure 3.3, which represents the negative feedback interconnection of a passive nonlinear system and an LTI system. First we introduce the following definition.

Definition: A system Σ is said to be *strongly* zero-state observable if it is zero-state observable, and has the property that $\{\lim_{t\to\infty} u(t) = 0$, and $\lim_{t\to\infty} y(t) = 0\}$ implies $\lim_{t\to\infty} x(t) = 0$.

Systems that are strongly zero-state observable constitute a large class of physical systems.

Theorems 3.7 and 3.8 establish the global asymptotic stability of the feedback system under weaker conditions, i.e., when one of the systems passive, and the other is linear and WSPR or MSPR.

Theorem 3.7 *Suppose in the system shown in Figure 3.3, suppose Σ is affine in control (see (3.14)), internally passive, strongly zero-state observable, and has a radially unbounded storage function $E(x)$. Then the feedback system is globally asyptotically stable if the system $G(s)$ is WSPR.*

Proof 3.7 Let x denote the n-dimensional state vector of Σ and $\hat{x}$ denote the $\hat{n}$-dimensional state vector of a minimal realization $[A, B, C, D]$ of $G(s)$. Since

$G(s)$ is WSPR, from [Loz.90], there exist matrices $P = P^T > 0,\ P \in \Re^{\hat{n}\times\hat{n}}, L \in \Re^{m\times\hat{n}}, W \in \Re^{m\times m}$, such that

$$A^T P + PA = -L^T L \tag{3.54}$$

$$C = B^T P + W^T L \tag{3.55}$$

$$W^T W = D + D^T \tag{3.56}$$

where $[A, B, L, W]$ is minimal and $F(s) = W + L(sI - A)^{-1}B$ is minimum-phase.

Consider the candidate Lyapunov function

$$V(x, \hat{x}) = E(x) + \hat{E}(\hat{x})$$

where $\hat{E}(\hat{x}) = \frac{1}{2}\hat{x}^T P \hat{x}$ is the storage function of $G(s)$. Note that $E(x)$ is positive definite because the system is zero-state observable [Hil.94]. Using (3.17) and proceeding as in the proof of Theorem 3.4, the time derivative of V can be obtained as

$$\dot{V} = -d(x, u) + y^T u - z^T z + \hat{y}^T \hat{u} \tag{3.57}$$

where

$$d(x, u) = [l(x) + W(x)]^T [l(x) + W(x)u]$$

and

$$z(t) = \frac{1}{\sqrt{2}}[L\hat{x} + W\hat{u}].$$

Noting that $u = -\hat{y}$, $\hat{u} = y$, we have

$$\dot{V} = -d(x, u) - z^T z \le 0 \tag{3.58}$$

i.e., $\dot{V}$ is negative semi-definite, and the closed-loop system is at least Lyapunov-stable. However, we will show that the closed-loop system is, in fact, globally asymptotically stable.

We have established so far that $\dot{V} \le 0$. Now, $\dot{V} \equiv 0 \Rightarrow d(x(t), u(t)) \equiv 0$, and $z(t) \equiv 0$. However, $z(t)$ is the output (produced by the input $\hat{u}(t)$) of the system: $\hat{F}(s) = \frac{1}{\sqrt{2}}[W + L(sI - A)^{-1}B]$, which has transmission zeros only in

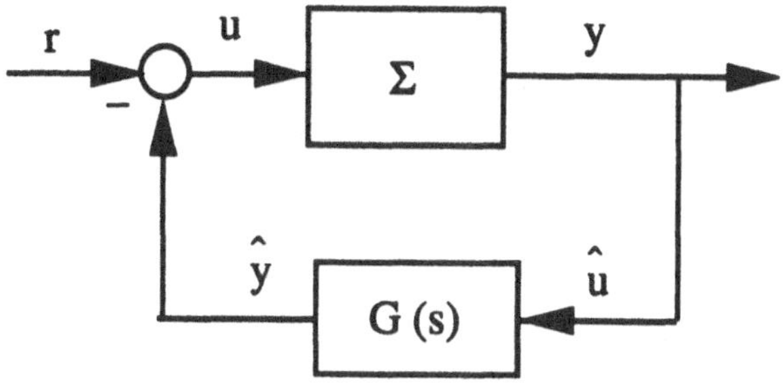

Figure 3.3: Feedback interconnection of passive nonlinear system and linear PR controller

the open left-half-plane. Therefore, $\hat{u}(t) \to 0$ exponentially, i.e., $y(t) \to 0$. Since $G(s)$ is stable, $\hat{u}(t) \to 0 \Rightarrow \hat{y}(t) \to 0$. That is, $u(t) \to 0$ and $y(t) \to 0$. Since Σ is strongly zero-state observable and $[A, B, C, D]$ is minimal, $u(t)$, $y(t) \to 0 \Rightarrow x(t) \to 0$, and $\hat{x} \to 0$, i.e., $\|x(t)\| \to 0$, $\|\hat{x}(t)\| \to 0$. This implies that $V(x(t), \hat{x}(t)) \to 0$, i.e., it shows that $V(x(t), \hat{x}(t))$ *decreases* ($\dot{V} < 0$) somewhere along the trajectories implying that $\dot{V} \not\equiv 0$. Therefore, such trajectories cannot exist, and by LaSalle's invariance principle, the feedback system is globally asymptotically stable. □

The conditions of Theorem 3.7 can be relaxed to allow the system $G(s)$ to have poles on the imaginary axis. However, this requires the nonlinear system Σ to satisfy an additional condition. The following theorem gives the conditions under which the feedback interconnection of a passive system and an MSPR system is globally asymptotically stable.

Theorem 3.8 *Consider the feedback system of Figure 3.3. Suppose Σ is affine in control, internally passive, strongly zero-state-observable, and has a radially unbounded storage function $E(x)$. In addition, suppose Σ has the property that $u(t) \notin \mathcal{L}_2[0, \infty) \Rightarrow \lim_{t\to\infty} y(t) \neq 0$. Then the closed-loop system is globally asymptotically stable if $G(s)$ is MSPR.*

Proof 3.8 Proceeding as in the proof of Theorem 3.7, suppose x denotes the n-dimensional state vector of Σ and $\hat{x}$ denote the $\hat{n}$-dimensional state vector of a minimal realization $[A, B, C, D]$ of $G(s)$. Let n_2 denote the number of poles of the stable part $G_2(s)$ of $G(s)$, and let $[A_2, B_2, C_2, D]$ denote its minimal realization. Since $G(s)$ is MSPR, from Lemma 2 of Sec. 3.3.3, there exist matrices $P = P^T > 0$, $P \in \Re^{n \times n}, \mathcal{L} \in \Re^{m \times n_2}, W \in \Re^{m \times m}$, such that Eqs. (3.27)-(3.30) are satisfied, and $[A_2, B_2, \mathcal{L}, W]$ is minimal and minimum-phase.

Consider the candidate Lyapunov function

$$V(x, \hat{x}) = E(x) + \hat{E}(\hat{x})$$

where $\hat{E}(\hat{x}) = \frac{1}{2}\hat{x}^T P \hat{x}$ is the storage function of $G(s)$. Using (3.17) and proceeding as in the proof of Theorem 7, the time derivative of V can be obtained as

$$\dot{V} = -d(x, u) + y^T u - z^T z + \hat{y}^T \hat{u} \tag{3.59}$$

where

$$d(x, u) = [l(x) + W(x)u]^T [l(x) + W(x)u]$$

and

$$z(t) = \frac{1}{\sqrt{2}}[L\hat{x} + W\hat{u}] = \frac{1}{\sqrt{2}}[\mathcal{L}\hat{x}_2 + W\hat{u}]$$

where $\hat{x}_2$ is the state vector corresponding to $[A_2, B_2, C_2, D]$. Noting that $u = -\hat{y}$, $\hat{u} = y$, we have

$$\dot{V} = -d(x, u) - z^T z \leq 0 \tag{3.60}$$

i.e., $\dot{V}$ is negative semi-definite, and the closed-loop system is at least Lyapunov-stable. However, we will show that the closed-loop system is, in fact, globally asymptotically stable.

$\dot{V} \equiv 0 \Rightarrow d(x(t), u(t)) \equiv 0$, and $z(t) \equiv 0$. However, $z(t)$ is the output (produced by the input $\hat{u}(t)$) of the system: $\hat{F}(s) = \frac{1}{\sqrt{2}}[W + \mathcal{L}(sI - A_2)^{-1}B_2]$, whose transmission zeros are in the open left-half-plane. Therefore, $\hat{u}(t) \to 0$ exponentially, i.e., $y(t) \to 0$. Therefore, $\hat{y}(t)$ can consist only of exponentially decaying terms, and sinusouidal terms (including zero-frequency) corresponding to the $j\omega$-axis poles of $G(s)$. If $\hat{y}(t)$ consists of any sinusoids, then $\hat{y}(t) \notin \mathcal{L}_2[0, \infty)$, which implies that $y(t)$ cannot go to zero; this contradicts the fact that $y(t) \to 0$; therefore, $y(t)$ can consist only of exponentially decaying terms. Since Σ is strongly zero-state observable and $[A, B, C, D]$ is minimal, $u(t)$, $y(t) \to 0 \Rightarrow x(t) \to 0$, and $\hat{x}(t) \to 0$. The rest of the proof is the same as that of Theorem 3.7. □

In Theorem 3.8, the requirement that $u(t) \notin \mathcal{L}_2[0, \infty) \Rightarrow \lim_{t \to 0} y(t) \neq 0$ will usually be satisfied if the zero dynamics of Σ are asymptotically stable.

Theorems 3.7 and 3.8 show that a class of nonlinear passive systems can be stabilized by a class of FDLTI passive controllers. The stability does not depend on the system order or the knowledge of the system parameters, and is therefore *robust*.

Chapter 4

Passivity-Based Controllers

4.1 Introduction

We consider the problem of controlling a class of multibody flexible spacecraft, such as space platforms with multiple articulated payloads and space-based manipulators used for satellite assembly and servicing. Such systems can have significant flexibility in their structural members as well as joints. Controller design for such systems is a difficult problem because of the highly nonlinear dynamics, large number of significant elastic modes with low inherent damping, and uncertainties in the mathematical model. The published literature contains a number of important stability results for some subclasses of this problem (e.g., linear flexible structures, nonlinear multibody rigid structures, and most recently, multibody flexible structures). Under certain conditions, the input-output maps for such systems can be shown to be passive. The Lyapunov and passivity approaches are used in [Jos.89] to prove global asymptotic stability of linear single-body flexible space structures (without articulated appendages) using a class of passivity-based controllers. These controllers were referred to as "dissipative controllers" because they result in dissipation of the systems' energy. (We shall use the term "dissipative controllers" to denote this class of passivity-based controllers). The stability properties were shown to be robust to first-order actuator dynamics and certain actuator-sensor nonlinearities. Multibody rigid structures represent another class of systems for which stability results have been advanced. Subject to a few restrictions, these systems can be ideally categorized as natural systems (see [Mei.70]). Such systems are known to exhibit global asymptotic stability under proportional-and-derivative (PD) control. After recognition that rigid manipulators belong to the class of natural systems, a number of researchers (e.g., [Tak.81, Kod.84]) established global asymptotic stability of terrestrial rigid manipulators using PD control with gravity compensation. Stability of tracking controllers was investigated in references [Wen.88] and [Pad.88] for rigid manipulators. Modeling and control of *flexible* manipulators has also been addressed in the literature (e.g., see [Boo.90], [Pad.90], and [Yua.93]). Lyapunov stability of multilink flexible systems was addressed in [Jua.93]. This chapter addresses global asymptotic stability of nonlinear, multilink, flexible space structures. The results presented are closely related to [Jos.95], [Kel.95, Kel.95a], and [Jos.95a].

The system considered is a complete nonlinear rotational dynamic model of a multibody flexible spacecraft derived in Chapter 2, which is assumed to have a branched geometry, (i.e., a central flexible body with various flexible appendage bodies. See Figure 2.1). The actuators and sensors are assumed to be compatible and collocated. Global asymptotic stability of such systems controlled by a nonlinear dissipative controller is investigated in this chapter. In many applications, the central body has large mass and moments of inertia when compared with that of the appendage bodies. As a result, the motion of the central body is small and can be considered to be in the linear range. For this case, robust stability is proved with linear static as well as with linear dynamic dissipative compensators. The effects of realistic nonlinearities in the actuators and sensors are also investigated. The stability proofs use Lyapunov's method and LaSalle's invariance principle. For systems with linear collocated actuators and sensors, the stability proof by Lyapunov's method can take a simpler form if the work-energy rate principle [Jua.93] is used. The work-energy rate principle is applicable only when the system is holonomic and scleronomic in nature. A more direct approach is used here in evaluating the time derivative of the Lyapunov function which makes the results more general.

The organization of the chapter is as follows. The nonlinear mathematical model of a generic flexible multibody system, which was presented in Chapter 2, is further investigated in section 4.2 and is shown to be internally passive. Some basic kinematic relations of the quaternion (i.e., measure of attitude of the central body) are also given. Section 4.3 establishes the global asymptotic stability of the complete nonlinear system under a nonlinear control law based on quaternion feedback. A special case, in which the central body attitude motion is small, is considered in section 4.4. Global stability is proved under static dissipative compensation, and these results are extended to the case in which certain actuator and sensor nonlinearities are present. In addition, dynamic dissipative compensators, a more versatile class of dissipative compensators, are considered in section 4.5. Section 4.6 presents the application of the results to an important special case, namely linear single-body flexible spacecraft. Numerical examples are given in section 4.7 to validate some of the results presented in preceding sections.

4.2 Some Properties of System Model

4.2.1 Equations of Motion

The class of systems considered consists of a branched configuration of flexible bodies as shown in Figure 2.1, where each branch by itself could be a serial chain of structures. For the sake of simplicity and without loss of generality, we consider a spacecraft with only one such branch where each appendage body has one degree of freedom (hinge) with respect to the previous body in the chain. However, the results presented are also applicable to the general case with multiple branches. The spacecraft is assumed to consist of a central flexible body and a chain of $(k-3)$ flexible links. The central body has three rigid rotational degrees of freedom, and each link is connected by one rotational degree of freedom to the neighboring link. Using Lagrangian formulation, the following equations of motion can be obtained (see Chapter 2).

$$M(p)\ddot{p} + C(p,\dot{p})\dot{p} + D\dot{p} + Kp = B^T u \tag{4.1}$$

where $\dot{p} = (\omega^T, \dot{\theta}^T, \dot{q}^T)^T$, ω is the 3×1 inertial angular velocity vector (in body-fixed coordinates) for the central body, $\theta = (\theta_1, \theta_2, \ldots, \theta_{(k-3)})^T$ (θ_i denotes the joint angle for the ith joint expressed in body-fixed coordinates), q is the $(n-k)$th vector of flexible degrees of freedom (modal amplitudes), $M(p) = M^T(p) > 0$ is the configuration-dependent mass-inertia matrix. $p = (\gamma^T, \theta^T, q^T)^T$, $\dot{\gamma} = \omega$, and $C(p,\dot{p})$ corresponds to Coriolis and centrifugal forces. $B = [I_{k\times k} \quad 0_{k\times(n-k)}]$ is the control influence matrix, and u is the k vector of applied torques. The first three components of u represent the attitude control torques applied to the central body about its X-, Y-, and Z-axes; the remaining components are the torques applied at the $(k-3)$ joints. The symmetric positive semidefinite stiffness and damping matrices K and D are given by

$$K = \begin{bmatrix} 0_{k\times k} & 0_{k\times(n-k)} \\ 0_{(n-k)\times k} & \tilde{K}_{(n-k)\times(n-k)} \end{bmatrix} \qquad D = \begin{bmatrix} 0_{k\times k} & 0_{k\times(n-k)} \\ 0_{(n-k)\times k} & \tilde{D}_{(n-k)\times(n-k)} \end{bmatrix} \tag{4.2}$$

where $\tilde{K}$ and $\tilde{D}$ are symmetric matrices corresponding to the flexible degrees of freedom. $\tilde{K}$ is inherently positive definite, and $\tilde{D}$ is assumed to be positive-definite. This assumption is realistic because all physical systems have some inherent damping.

The angular measurements for the central body are Euler angles (not the vector γ), whereas the remaining angular measurements between bodies are relative angles. An important inherent property of such systems that is crucial to the stability results was proved in Chapter 2 and is re-stated next.

Property S: For the system in Eq. (4.1), the matrix $(\frac{1}{2}\dot{M} - C)$ is skew-symmetric $\forall\, t$.

The central-body attitude (Euler angle) vector η is given by $E(\eta)\dot{\eta} = \omega$, where $E(\eta)$ is a 3×3 transformation matrix. (See [Gre.88].) The sensor outputs consist of three central-body Euler angles, the $(k-3)$ joint angles, and the angular rates (i.e., the sensors are collocated with the torque actuators). The sensor outputs are then given by

$$y_p = B\widehat{p} \qquad y_r = B\dot{p} \tag{4.3}$$

where $\widehat{p} = (\eta^T, \theta^T, q^T)^T$ and η is the Euler angle vector for the central body. The measured angular position and rate vectors are $y_p = (\eta^T, \theta^T)^T$ and $y_r = (\omega^T, \dot{\theta}^T)^T$, respectively. The body rate measurements ω are assumed to be available from rate gyros. The system in (4.1), with state vector $\overline{x} = (p^T, \dot{p}^T)^T$ is affine in control and is internally passive (with respect to output y_r), as shown next.

Theorem 4.1 *The system in (4.1) (with output y_r) is internally passive, i.e., there exists a storage function $E(\overline{x}) \geq 0$ such that*

$$\int_0^T y_r^T(t)u(t) \geq E[\overline{x}(T)] - E[\overline{x}(0)] \ \ \forall\, T \geq 0, \ \forall\, u \in \mathcal{L}_{2e}^k. \tag{4.4}$$

Proof 4.1 Pre-multiplying both sides of Eq. (4.1) by $\dot{p}^T$ and integrating, we get

$$\int_0^T [\dot{p}^T M(p)\ddot{p} + \dot{p}^T C(p,\dot{p})\dot{p} + \dot{p}^T D\dot{p} + \dot{p}^T Kp] = \int_0^T y_r^T u dt.$$

Noting that

$$\frac{d}{dt}[\dot{p}^T M(p)\dot{p}] = 2\dot{p}^T M(p)\ddot{p} + \dot{p}^T \dot{M}(p)\dot{p}$$

and using Property $\mathcal{S}$,

$$\frac{1}{2}[\dot{p}^T(T)M(p(T))\dot{p}(T) + p^T(T)Kp(T)] - \frac{1}{2}[\dot{p}^T(0)M(p(0))\dot{p}(0) + p^T(0)Kp(0)] + \int_0^T \dot{p}^T D\dot{p} dt = \int_0^T y_r^T u dt. \tag{4.5}$$

That is,

$$< y_r, u >_T \ \geq \ E[\overline{x}(T)] - E[\overline{x}(0)] \tag{4.6}$$

with $E(\overline{x}) = \frac{1}{2}[\dot{p}^T M(p)\dot{p} + p^T Kp]$ □

Since the system is internally passive, using Theorem 3.2, there exist functions $l(\overline{x})$, $W(\overline{x})$ such that Eq. (3.18) holds. Note that $N(\overline{x}) = 0$ for this system and therefore $W(\overline{x}) = 0$. It can be verified that

$$\frac{d}{dt}E[\overline{x}(t)] = -d(\overline{x}) + y_r^T u$$

where $d(\overline{x}) = \dot{p}^T D\dot{p}$. Therefore, $l(\overline{x}) = D^{\frac{1}{2}}\dot{p}$ (See (3.17)). Note that $E(\overline{x})$ is only positive semidefinite because K is positive semidefinite.

4.2.2 Quaternion as Measure of Attitude

The orientation of a free-floating body can be minimally represented by a three-dimensional orientation vector. However, this representation is not unique. Euler angles are commonly used as a minimal representation of the attitude. As stated previously, the 3×1 Euler angle vector η is given by $E(\eta)\dot{\eta} = \omega$, where $E(\eta)$ is a 3×3 transformation matrix. For certain values of η, $E(\eta)$ becomes singular; however, note that the limitations imposed on the allowable orientations because of this singularity are purely mathematical in nature and are not restrictive in the physical sense. The problem of singularity in the three-parameter representation of attitude has been studied in detail in the literature. An effective way of overcoming the singularity problem is to use the quaternion formulation. (See [Wen.91, Kan.73, Ick.70, Har.64].)

The unit quaternion (Euler parameter vector) α is defined as follows:

$$\alpha = (\overline{\alpha}^T, \alpha_4)^T \qquad \overline{\alpha} = \begin{bmatrix} \widehat{\alpha}_1 \\ \widehat{\alpha}_2 \\ \widehat{\alpha}_3 \end{bmatrix} \sin\left(\frac{\phi}{2}\right) \qquad \alpha_4 = \cos\left(\frac{\phi}{2}\right) \tag{4.7}$$

where $\widehat{\alpha} = (\widehat{\alpha}_1, \widehat{\alpha}_2, \widehat{\alpha}_3)^T$ is the unit vector along the eigenaxis of rotation and ϕ is the magnitude of rotation. The quaternion is subject to the norm constraint

$$\overline{\alpha}^T\overline{\alpha} + \alpha_4^2 = 1. \tag{4.8}$$

The quaternion can also be shown [Har.64] to obey the following kinematic differential equations:

$$\dot{\overline{\alpha}} = \frac{1}{2}(\omega \times \overline{\alpha} + \alpha_4\omega) \tag{4.9}$$

$$\dot{\alpha}_4 = -\frac{1}{2}\omega^T\overline{\alpha}. \tag{4.10}$$

The attitude control of a single-body rigid spacecraft with quaternion feedback has been thoroughly investigated in [Wen.91], [Kan.73], [Ick.70], and [Har.64]

The quaternion representation is used here for the central-body attitude. The quaternion can be computed from Euler angle measurements given by equations (4.3). (See [Hau.89])

The open-loop system, given by equations (4.1), (4.9), and (4.10) with $u = 0$, has multiple equilibrium solutions $(\overline{\alpha}_{ss}^T, \alpha_{4ss}, \theta_{ss}^T)^T$, where the subscript ss denotes the steady-state value; the steady-state value of q and $\dot{p}$ is zero. By defining $\beta = (\alpha_4 - 1)$ and denoting $\dot{p} = z$, equations (4.1), (4.9), and (4.10) can be rewritten as

$$M\dot{z} + Cz + Dz + K\left(0_{1\times 3}, \theta^T, q^T\right)^T = B^T u \tag{4.11}$$

$$\begin{bmatrix} \dot{\theta} \\ \dot{q} \end{bmatrix} = [0_{(n-3)\times 3} I_{n-3}]z \tag{4.12}$$

$$\dot{\overline{\alpha}} = \frac{1}{2}\left[\omega \times \overline{\alpha} + (\beta + 1)\omega\right] \tag{4.13}$$

$$\dot{\beta} = -\frac{1}{2}\omega^T\overline{\alpha}. \tag{4.14}$$

In equation (4.11) the matrices M and C are functions of p and $(p, \dot{p})$, respectively. Note that the first three elements of p associated with the orientation of the central body can be fully described by the unit quaternion. Hence, M and C are implicit functions of α, and therefore the system represented by equations (4.11)-(4.14) is autonomous and can be expressed in the state-space form as follows:

$$\dot{x} = f(x) + g(x)u \tag{4.15}$$

where $x = (\overline{\alpha}^T, \beta, \theta^T, q^T, z^T)^T$ and $g(x)$ is a $(2n + 1) \times k$ matrix. Note that the dimension of x is $(2n + 1)$, one more than the dimension of the system in equation (4.1). However, the constraint of equation (4.8) is now present. Verification that the constraint of equation (4.8) is satisfied for all $t > 0$ if it is satisfied at $t = 0$ easily follows from equations (4.9) and (4.10).

4.3 Nonlinear Dissipative Control Laws

Consider the control law u, given by:

$$u = -G_p\tilde{p} - G_r y_r \tag{4.16}$$

where $\tilde{p} = (\overline{\alpha}^T, \theta^T)^T$. Matrices G_p and G_r are symmetric positive definite $(k \times k)$ matrices and G_p is given by:

$$G_p = \mathrm{diag}\left[\frac{1}{2}\{(\tilde{\overline{\alpha}} + I_3\alpha_4)G_{p1} + \nu(1-\alpha_4)I_3\}, G_{p2}\right] \tag{4.17}$$

where ν is a positive scalar, $\tilde{\overline{\alpha}}$ represents a cross product operator matrix $(\overline{\alpha}\times)$, and G_{p1}, G_{p2} are (3×3) and $(k-3) \times (k-3)$ symmetric positive definite matrices, respectively. Note that the feedback $\tilde{p}$ can be computed using the sensor measurements y_p using quaternion equations (4.7). The following Lemma gives the closed-loop equilibrium solutions [Kel.95a]. Note that this is an extension of Lemma 1 in [Jos.95a] to multibody flexible spacecraft.

Lemma 1. *Suppose G_{p1} and G_{p2} are symmetric and positive definite, and $0 < \lambda_M(G_{p1}) \leq 2\nu$, where $\lambda_M(\cdot)$ denotes the largest eigenvalue. Then the closed-loop system given by (4.15) and (4.16) has exactly two equlibrium solutions: $[\tilde{p} = q = \dot{p} = 0, \beta = 0]$ and $[\tilde{p} = q = \dot{p} = 0, \beta = -2]$.*

Proof 1. The closed-loop equilibrium solutions can be obtained by equating all time derivatives in Eqs. (4.11)-(4.14) to zero, i.e., $\dot{p} = \ddot{p} = 0 \Rightarrow \omega = 0$, $\dot{\theta} = 0$, and $\dot{q} = 0$. Substituting in Eq. (4.11), we get

$$-B^T G_p\tilde{p} = \begin{bmatrix} -\frac{1}{2}[\{\tilde{\overline{\alpha}} + I_3(\beta+1)\}G_{p1} - \nu\beta I_3]\overline{\alpha} \\ -G_{p2}\theta \\ 0_{(n-k-3)\times 1} \end{bmatrix} = \begin{bmatrix} 0_{3\times 1} \\ 0_{k\times 1} \\ \tilde{K}q \end{bmatrix} \tag{4.18}$$

$$\Rightarrow \frac{1}{2}[\{\tilde{\overline{\alpha}} + I_3(\beta+1)\}G_{p1} - \nu\beta I_3]\overline{\alpha} = 0,\ \theta = 0,\ q = 0. \tag{4.19}$$

Premultiplying first of the three Eqs. in (4.19) by $\overline{\alpha}^T$, we get

$$\overline{\alpha}^T \mathcal{M}\overline{\alpha} = 0, \text{ where } \mathcal{M} = (\beta+1)G_{p1} - \nu\beta I_3. \tag{4.20}$$

The eigenvalues of $\mathcal{M}$ are given by: $\lambda_i(\mathcal{M}) = (\beta+1)\lambda_i(G_{p1}) - \nu\beta = (\beta + 1)(\lambda_i(G_{p1}) - \nu) + \nu$. It can be easily proved that for the cases when $0 <$

$\lambda_i(G_{p1}) < \nu, \lambda_i(G_{p1}) = \nu$, and $\nu < \lambda_i(G_{p1}) < 2\nu$, the matrix $\mathcal{M}$ is nonsingular for any feasible values of $\overline{\alpha}$. For $\lambda_i(G_{p1}) = 2\nu$, $\mathcal{M}$ is singular only when $\beta = -2$, i.e., when $\overline{\alpha} = 0$. For the case when $\lambda_i(G_{p1}) > 2\nu$, there are feasible nonzero solutions of $\overline{\alpha}$ for which $\mathcal{M}$ could become singular. Hence, for $\lambda_i(G_{p1}) \leq 2\nu$ closed-loop system has only two possible equilibrium solutions:$[\tilde{p} = 0,\ q = 0,\ \dot{p} = 0,\ \beta = 0]$ and $[\tilde{p} = 0\ q = 0\ \dot{p} = 0, \beta = -2]$.

Thus, there appear to be two closed-loop equilibrium points corresponding to $\beta = 0$ and $\beta = -2$ (all other state variables being zero). However, from equation (4.7), $\beta = 0 \Rightarrow \phi = 0$, and $\beta = -2 \Rightarrow \phi = 2l\pi$, i.e., there is only one equilibrium point in the physical space. One of the objectives of the control law is to transfer the state of the system from one orientation (i.e., equilibrium position) to another orientation. Without loss of generality, the target orientation can be defined to be zero, and the initial orientation, given by $(\overline{\alpha}(0), \alpha_4(0), \theta(0))$, can always be defined in such a way that $|\phi_i(0)| \leq \pi$, $0 \leq \alpha_4(0) \leq 1$ (corresponds to $|\phi| \leq \pi$), and $(\overline{\alpha}(0), \alpha_4(0))$ satisfy equation (4.8). □

The global asymptotic stability of the physical equilibrium state (the origin of the state-space) is proved next. (We shall use the phrase "global asymptotic stability of a system" to denote the global asymptotic stability of the origin).

Theorem 4.2 *Suppose G_{p1}, G_{p2}, and G_r are symmetric and positive definite, and $0 < \lambda_M(G_{p1}) \leq 2\nu$. Then the closed-loop system given by Eqs. (4.15), (4.16) and (4.17) is globally asymptotically stable.*

Proof 4.2 Consider the candidate Lyapunov function

$$V = \frac{1}{2}\dot{p}^T M(p)\dot{p} + \frac{1}{2}q^T \tilde{K} q + \frac{1}{2}\theta^T G_{p2}\theta + \frac{1}{2}\overline{\alpha}^T G_{p1}\overline{\alpha} + \frac{1}{2}\nu\beta^2. \tag{4.21}$$

V is a positive definite function of the state vector x, and is radially unbounded. Also note that the matrix $M(p)$, although configuration dependent, is uniformly bounded from below and above by the values which correspond to the minimum and maximum inertia configurations, respectively (i.e., there exist positive definite matrices $\underline{M}$ and $\overline{M}$ such that $\underline{M} \leq M \leq \overline{M}$). Taking time derivative of

V, evaluating $\dot{V}$ along system trajectories, and using Lemma 1 yields

$$\dot{V} = -\dot{q}^T \tilde{D}\dot{q} + (B\dot{p})^T u + \dot{\theta}^T G_{p2}\theta + \overline{\alpha}^T G_{p1}\dot{\overline{\alpha}} + \nu\beta\dot{\beta}. \tag{4.22}$$

Substituting Eqs.(4.13) and (4.14) in (4.22) and rearranging the terms, we get

$$\dot{V} = -\dot{q}^T \tilde{D}\dot{q} + (B\dot{p})^T u + \dot{\theta} G_{p2}\theta + \frac{\omega^T}{2}\left[\tilde{\alpha} G_{p1} + (\beta+1)G_{p1} - \nu\beta\right]\overline{\alpha}. \tag{4.23}$$

Using control law (4.16) in Eq.(4.23) yields

$$\dot{V} = -\dot{q}^T \tilde{D}\dot{q} - y_r^T G_r y_r = -\dot{p}^T (D + B^T G_r B)\dot{p}. \tag{4.24}$$

Since $(D + B^T G_r B)$ is a positive definite symmetric matrix, $\dot{V} \leq 0$, i.e., $\dot{V}$ is negative semidefinite. Also, $\dot{V} = 0 \Rightarrow \dot{p} = 0 \Rightarrow \ddot{p} = 0$. Substituting in the closed-loop equation and using Lemma 1 gives two equilibrium solutions: $[\tilde{p} = 0\ , q = 0,\ \dot{p} = 0, \beta = 0]$ and $[\tilde{p} = 0,\ q = 0,\ \dot{p} = 0, \beta = -2]$. However, as stated previously, these two equilibrium points correspond to the same physical equilibrium state [Jos.95a]. Hence, $\dot{V}$ is negative along all system trajectories except at the two equlibrium points which represent the same physical equilibrium state. From equation (4.21), it can be verified that any small perturbation ϵ in α_4 from the equilibrium point corresponding to $\alpha_4 = -1$ will cause a decrease in the value of V ($\epsilon > 0$ because $|\alpha_4| \leq 1$). Thus, in the mathematical sense, $\alpha_4 = -1$ corresponds to an isolated equilibrium point such that $\dot{V} = 0$ at that point and $\dot{V} < 0$ in the neighborhood of that point (i.e., $\alpha_4 = -1$ is a repeller and not an attractor). Previously, $\dot{V}$ has been shown to be negative along all trajectories in the state space except at the two equilibrium points. That is, if the system's initial condition lies anywhere in the state space except at the equilibrium point corresponding to $\alpha_4 = -1$, then the system state will asymptotically approach the origin (i.e., $x = 0$); if the system is at the equilibrium point corresponding to $\alpha_4 = -1$ at $t = 0$, then it will stay there for all $t > 0$. However, this is the same equilibrium point in the physical space; hence, by LaSalle's invariance principle, the system is globally asymptotically stable.

□

4.3.0.1 A Scalar-Gain Control Law

We next consider another nonlinear control law which uses a scalar gain for the quaternion.

Consider the control law u, given by (4.16) repeated below

$$u = -G_p\widetilde{p} - G_r y_r \tag{4.25}$$

where $\widetilde{p} = (\overline{\alpha}^T, \theta^T)^T$. Matrices G_p and G_r are symmetric positive definite $k \times k$ matrices; G_p is given by

$$G_p = \begin{bmatrix} (1 + \frac{(\beta+1)}{2})G_{p1} & 0_{3\times(k-3)} \\ 0_{(k-3)\times 3} & G_{p2_{(k-3)\times(k-3)}} \end{bmatrix}. \tag{4.26}$$

The closed-loop equilibrium solutions can be obtained by equating all the derivatives to zero in equations (4.1), (4.9), and (4.10). In particular, $\dot{p} = \ddot{p} = 0 \Rightarrow \omega = 0, \theta = 0, \dot{q} = 0$, and

$$-B^T G_p\widetilde{p} = \begin{bmatrix} -G_p\widetilde{p} \\ 0_{(n-k)\times(n-k)} \end{bmatrix} = \begin{bmatrix} 0_{k\times 1} \\ \widetilde{K}q \end{bmatrix}. \tag{4.27}$$

Because of equation (4.8), $|\beta + 1| \leq 1$. Therefore G_p is positive definite and equation (4.27) implies $\widetilde{p} = (\overline{\alpha}^T, \theta^T)^T = 0$ and $q = 0$. The equilibrium solution of equation (4.10) is $\beta = \beta_{ss} =$ Constant (i.e., $\alpha_4 =$ Constant), which implies from equation (4.8) that $\alpha_4 = \pm 1$ (because $\overline{\alpha} = 0$). However, the two equilibrium solutions correspond to the same equilibrium in the physical space. Unlike the control law of Eqs. (4.16) and (4.17), there are no restrictions on the gains. In the case of a rest-to-rest maneuver, the target orientation can be defined to be zero without loss of generality.

The following theorem establishes the global asymptotic stability of the physical equilibrium state of the system.

Theorem 4.3 *Suppose $G_{p2_{(k-3)\times(k-3)}}$ and $G_{r_{k\times k}}$ are symmetric and positive definite, and $G_{p1} = \mu I_3$, where $\mu > 0$. Then, the closed-loop system given by equations (4.15), (4.25) and (4.26) is globally asymptotically stable.*

Proof 4.3 Consider the candidate Lyapunov function

$$V = \frac{1}{2}\dot{p}^T M(p)\dot{p} + \frac{1}{2}q^T\widetilde{K}q + \frac{1}{2}\theta^T G_{p2}\theta + \frac{1}{2}\overline{\alpha}^T(G_{p1} + 2\mu I_3)\overline{\alpha} + \mu\beta^2. \tag{4.28}$$

Here, V is clearly positive definite and radially unbounded with respect to a state vector $(\overline{\alpha}^T, \beta, \theta^T, q^T, \dot{p}^T)^T$ because $M(p)$, $\widetilde{K}$, G_{p1}, and G_{p2} are all positive definite symmetric matrices. The time derivative of V results in

$$\dot{V} = \dot{p}^T M\ddot{p} + \frac{1}{2}\dot{p}^T \dot{M}\dot{p} + \dot{q}^T \widetilde{K}q + \dot{\theta}^T G_{p2}\theta + \dot{\overline{\alpha}}^T (G_{p1} + 2\mu I_3)\overline{\alpha} + 2\mu\beta\dot{\beta}. \quad (4.29)$$

With the use of equations (4.1), (4.3), (4.13), (4.14), and (4.26),

$$\begin{aligned}\dot{V} &= \dot{p}^T B^T u + \dot{p}^T \left(\frac{1}{2}\dot{M} - C\right)\dot{p} - \dot{p}^T D\dot{p} - \dot{p}^T Kp + \dot{q}^T \widetilde{K}q + \dot{\theta}^T G_{p2}\theta \\ &\quad + \frac{1}{2}(\Omega\overline{\alpha})^T G_{p1}\overline{\alpha} + \frac{1}{2}(\beta+1)\omega^T G_{p1}\overline{\alpha} + \mu\omega^T\overline{\alpha} \qquad (4.30)\end{aligned}$$

where $\Omega = (\omega\times)$ denotes the skew-symmetric cross product matrix (i.e.,$\omega \times x = \Omega x$). With the substitution for u, the fact that $\dot{p}^T Kp = \dot{q}^T \widetilde{K}q$ and $(\Omega\overline{\alpha})^T G_{p1}\overline{\alpha} = 0$, and the use of Property $\mathcal{S}$ of the system, equation (4.30) becomes

$$\dot{V} = -\dot{p}^T (D + B^T G_r B)\dot{p} - (B\dot{p})^T G_p\widetilde{p} + \frac{1}{2}(\beta+1)\omega^T G_{p1}\overline{\alpha} + \mu\omega^T\overline{\alpha} + \dot{\theta}^T G_{p2}\theta. \quad (4.31)$$

Note that $(B\dot{p})^T G_p\widetilde{p} = \frac{1}{2}(\beta+1)\omega^T G_{p1}\overline{\alpha} + \mu\omega^T\overline{\alpha} + \dot{\theta}^T G_{p2}\theta$. After several cancellations,

$$\dot{V} = -\dot{p}^T (D + B^T G_r B)\dot{p}. \quad (4.32)$$

Because $(D + B^T G_r B)$ is a positive definite symmetric matrix, $\dot{V} \leq 0$ (i.e., $\dot{V}$ is negative semidefinite) and $\dot{V} = 0 \Rightarrow \dot{p} = 0 \Rightarrow \ddot{p} = 0$. By substitution in the closed-loop equation, Eq. (4.27) results. As shown previously, Eq. (4.27) $\Rightarrow \widetilde{p} = 0$ and $q = 0$; i.e., $\overline{\alpha} = 0$, $\theta = 0$, and $\alpha_4 = \pm 1$ (or $\beta = 0$ or -1). Consistent with the previous discussion, these values correspond to two equilibrium points representing the same physical equilibrium state. Therefore, by LaSalle's invariance principle (as used in Proof 4.2) it can be concluded that the system is globally asymptotically stable. □

The control gain of Eq. (4.26) can be shown to be a special case of the control gain of Eq. (4.17) obtained by setting $G_{p1} = 3\mu I_3$ and $\nu = 2\mu$. There are no restrictions on μ, as long as it is positive.

4.4 Systems in Attitude-Hold Configuration

Consider a special case where the central-body attitude motion is small. This can occur in many real situations. For example, in cases of space station-based or shuttle-based manipulators, the inertia of the base (central body) is much greater than that of any manipulator link or payload. In such cases, the rotational motion of the base can be assumed to be in the linear region, although the payloads (or links) attached to it can undergo large rotational and translational motions and nonlinear dynamic loading because of Coriolis and centripetal accelerations. The attitude of the central body is simply γ (the integral of the inertial angular velocity ω) and the use of quaternions is not necessary. The equations of motion (4.1) can now be expressed in the state-space form simply as

$$\dot{\overline{x}} = \begin{bmatrix} 0 & I \\ -M^{-1}K & -M^{-1}(C+D) \end{bmatrix} \overline{x} + \begin{bmatrix} 0 \\ B^T \end{bmatrix} u \tag{4.33}$$

where $\overline{x} = (p^T, \dot{p}^T)^T$, $p = (\gamma^T, \theta^T, q^T)^T$, and M and C are functions of $\overline{x}$.

4.4.1 Stability With Static Dissipative Controllers

The static dissipative control law u is given by

$$u = -\overline{G}_p y_p - G_r y_r \tag{4.34}$$

where $\overline{G}_p$ is symmetric positive definite $k \times k$ matrix and

$$y_p = Bp \qquad y_r = B\dot{p} \tag{4.35}$$

where y_p and y_r are measured angular position and rate vectors, respectively.

Theorem 4.4 *Suppose $\overline{G}_{p_{k\times k}}$ and $G_{r_{k\times k}}$ are symmetric and positive definite. Then, the closed-loop system given by equations (4.33)-(4.35) is globally asymptotically stable.*

Proof 4.4 Consider the candidate Lyapunov function

$$V = \frac{1}{2}\dot{p}^T M(p)\dot{p} + \frac{1}{2}p^T\left(K + B^T\overline{G}_p B\right)p. \tag{4.36}$$

Clearly, V is positive definite because $M(p)$ and $(K + B^T\overline{G}_pB)$ are positive definite symmetric matrices. Defining, $\overline{K} = (K + B^T\overline{G}_pB)$, the time derivative of V can be shown to be

$$\dot{V} = \dot{p}^T\left(\frac{1}{2}\dot{M} - C\right)\dot{p} - \dot{p}^T\overline{K}p + \dot{p}^T\overline{K}p - \dot{p}^T\left(D + B^TG_rB\right)\dot{p}. \tag{4.37}$$

Again, with the use of property $\mathcal{S}$, $\dot{p}^T(\frac{1}{2}\dot{M} - C)\dot{p} = 0$, and after some cancellations,

$$\dot{V} = -\dot{p}^T\left(D + B^TG_rB\right)\dot{p}. \tag{4.38}$$

Because $(D + B^TG_rB)$ is a positive definite symmetric matrix, $\dot{V} \leq 0$ (i.e., $\dot{V}$ is negative semidefinite in p and $\dot{p}$ and $\dot{V} = 0 \Rightarrow \dot{p} = 0 \Rightarrow \ddot{p} = 0$). With substitution in the closed-loop equation (4.1) where u is given by equation (4.34),

$$\left(K + B^T\overline{G}_pB\right)p = 0 \qquad \Rightarrow p = 0. \tag{4.39}$$

Thus, $\dot{V}$ is not zero along any trajectories; then, by LaSalle's invariance principle, the system is globally asymptotically stable. □

The significance of the results presented in Theorems 4.2-4.4 is that any nonlinear multibody system belonging to these classes can be robustly stabilized with the dissipative control laws given. In the case of manipulators, this means that any terminal angular position can be achieved from any initial position with guaranteed asymptotic stability.

4.4.2 Robustness to Actuator-Sensor Nonlinearities

Theorem 4.4, which assumes linear actuators and sensors, proves global asymptotic stability for systems in the attitude-hold configuration. In practice, however, the actuators and sensors have nonlinearities. The following theorem extends the results of [Jos.89] to the case of nonlinear flexible multibody systems. That is, the robust stability property of the dissipative controller is proved to hold in the presence of a broad class of actuator-sensor nonlinearities with the following definition: a function $\psi(\sigma)$ is said to belong to the $(0,\infty)$ sector (Figure 4.1(a)) if $\psi(0) = 0$ and $\sigma\psi(\sigma) > 0$ for $\sigma \neq 0$ and ψ is said to belong to the $[0,\infty)$ sector if $\sigma\psi(\sigma) \geq 0$.

Let $\psi_{ai}(\cdot)$, $\psi_{pi}(\cdot)$, and $\psi_{ri}(\cdot)$ denote the nonlinearities in the ith actuator, position sensor, and rate sensor channels, respectively. Both $\overline{G}_p$ and G_r are assumed to be diagonal with elements $\overline{G}_{pi}$ and $\overline{G}_{ri}$, respectively; then the actual input is given by

$$u_i = \psi_{ai}\left[-\overline{G}_{pi}\psi_{pi}\left(y_{pi}\right) - G_{ri}\psi_{ri}\left(y_{ri}\right)\right] \quad (i = 1, 2, .., k). \tag{4.40}$$

With the assumption that ψ_{pi}, ψ_{ai}, and $\psi_{ri}\,(i = 1, 2, \ldots, k)$ are continuous single-valued functions, $\Re \to \Re$ and the following theorem gives sufficient conditions for stability.

Theorem 4.5 *Consider the closed-loop system given by equations (4.33), (4.35), and (4.40), where $\overline{G}_p$ and G_r are diagonal with positive entries. Suppose ψ_{ai}, ψ_{pi}, and ψ_{ri} are single-valued, time-invariant continuous functions, and that (for $i = 1, 2, \ldots, k$)*

1. ψ_{ai} are monotonically nondecreasing and belong to the $(0, \infty)$ sector

2. ψ_{pi} and ψ_{ri} belong to the $(0, \infty)$ sector.

Under these conditions, the closed-loop system is globally asymptotically stable.

Proof 4.5 The proof closely follows that of [Jos.89], which addressed the linear, single-body spacecraft case. Let $\varphi = -y_p$ (k vector). Define

$$\overline{\psi}_{pi}\left(\sigma\right) = -\psi_{pi}\left(-\sigma\right) \tag{4.41}$$

$$\overline{\psi}_{ri}\left(\sigma\right) = -\psi_{ri}\left(-\sigma\right). \tag{4.42}$$

If ψ_{pi}, $\psi_{ri} \in (0, \infty)$ or $[0, \infty)$ sector, then $\overline{\psi}_{pi}$, $\overline{\psi}_{ri}$ also belong to the same sector. Now, consider the following Luré-Postnikov Lyapunov function:

$$V = \frac{1}{2}\dot{p}^T M\left(p\right)\dot{p} + \frac{1}{2}q^T \widetilde{K} q + \sum_{i=1}^{k} \int_0^{\varphi_i} \psi_{ai}\left[\overline{G}_{pi}\overline{\psi}_{pi}\left(\sigma\right)\right] d\sigma \tag{4.43}$$

where $\widetilde{K}$ is the symmetric positive definite part of K. Differentiation with respect to t and use of equation (4.1) yield

$$\dot{V} = \dot{p}^T\left(B^T u - C\dot{p} - D\dot{p} - Kp\right) + \frac{1}{2}\dot{p}^T \dot{M}\dot{p} + \sum_{i=1}^{k}\dot{\varphi}_i\psi_{ai}\left[\overline{G}_{pi}\overline{\psi}_{pi}\left(\varphi_i\right)\right] + \dot{q}^T\widetilde{K}q. \tag{4.44}$$

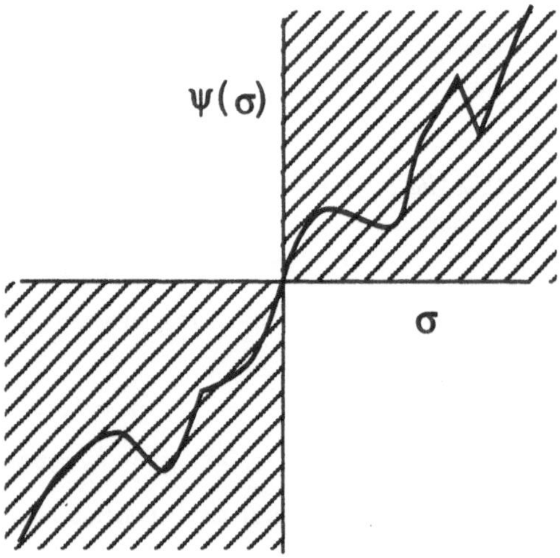

(a) (0, ∞) - sector nonlinearity

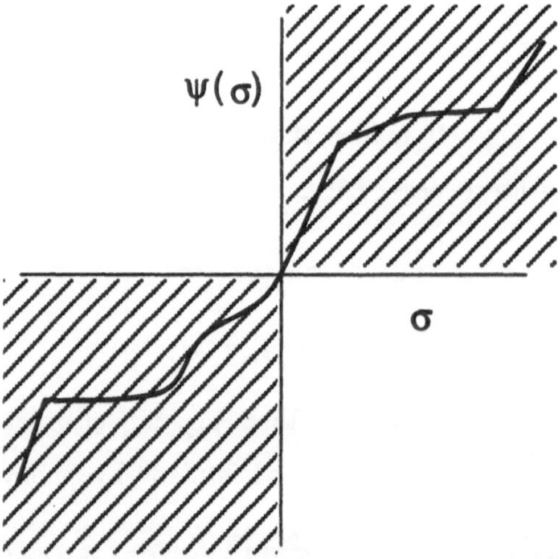

(b) (0,∞) - sector monotonically nondecreasing nonlinearity

Figure 4.1: Nonlinearities belonging to $(0, \infty)$-sector

Upon several cancellations and the use of property $\mathcal{S}$,

$$\dot{V} = \sum_{i=1}^{k} u_i y_{ri} - \dot{q}^T \widetilde{D} \dot{q} + \sum_{i=1}^{k} \dot{\varphi}_i \psi_{ai} \left[\overline{G}_{pi} \overline{\psi}_{pi} (\varphi_i) \right] \tag{4.45}$$

where matrix $\widetilde{D}$ is the positive definite part of D.

$$\dot{V} = -\dot{q}^T \widetilde{D} \dot{q} - \sum_{i=1}^{k} \dot{\varphi}_i \left\{ \psi_{ai} \left[G_{ri} \overline{\psi}_{ri} (\dot{\varphi}_i) + \overline{G}_{pi} \overline{\psi}_{pi} (\varphi_i) \right] - \psi_{ai} \left[\overline{G}_{pi} \overline{\psi}_{pi} (\varphi_i) \right] \right\}. \tag{4.46}$$

Because ψ_{ai} are monotonically nondecreasing and ψ_{ri} belong to the $(0, \infty)$ sector, $\dot{V} \leq 0$, and the system is at least Lyapunov-stable. In fact, it will be proved next that the system is globally asymptotically stable. First, consider a special case when ψ_{ai} are monotonically increasing. Then, $\dot{V} \leq -\dot{q}^T \widetilde{D} \dot{q}$, and $\dot{V} = 0$ only when $\dot{q} = 0$ and $\dot{\varphi} = 0$, which implies $\dot{p} = 0 \Rightarrow \ddot{p} = 0$. Substitution in the closed-loop equation results in

$$Kp = B^T \psi_a \left[-\overline{G}_p \psi_p (y_p) \right] \tag{4.47}$$

$$\begin{bmatrix} 0 \\ \widetilde{K} q \end{bmatrix} = \begin{bmatrix} \psi_a \left[-\overline{G}_p \psi_p (y_p) \right] \\ 0 \end{bmatrix} \tag{4.48}$$

$$\Rightarrow \psi_a \left[-\overline{G}_p \psi_p (y_p) \right] = 0 \quad \text{and} \quad q = 0. \tag{4.49}$$

If ψ_{pi} belong to the $(0, \infty)$ sector, $\psi_{ai}(\sigma) = \psi_{pi}(\sigma) = 0$ only when $\sigma = 0$. Therefore, $y_p = 0$. Thus, $\dot{V} = 0$ only at the origin, and the system is globally asymptotically stable.

In the case when actuator nonlinearities are of the monotonically nondecreasing type (e.g., saturation nonlinearity), $\dot{V}$ can be 0 even if $\dot{\varphi} \neq 0$. Figure 4.1(b) shows the monotonically nondecreasing type of nonlinearity. However, every system trajectory along which $\dot{V} \equiv 0$ will be shown to go to the origin asymptotically. When $\dot{\varphi} \neq 0$, $\dot{V} \equiv 0$ only when all actuators are saturated. Then, from the equations of motion, the system trajectories will go unbounded, which is not possible because the system was already proved to be Lyapunov-stable. Hence, $\dot{V}$ cannot be identically zero along the system trajectories, and the system is globally asymptotically stable. □

For the case when the central-body motion is not in the linear range, the results of robust stability in the presence of actuator-sensor nonlinearities cannot be easily extended because the stabilizing control laws are nonlinear.

The next subsection extends the robust stability results to a class of more versatile controllers called dynamic dissipative controllers. The advantages of using dynamic dissipative controllers include higher performance, more design freedom, and better noise attenuation.

4.4.3 Stability With Dynamic Dissipative Controllers

The performance achievable by the static dissipative control law is inherently limited because of its simple structure which permits direct input of noise into the system. To obtain better performance without the loss of guaranteed robustness to unmodeled dynamics and parametric uncertainties, we consider a class of dynamic dissipative controllers (DDC). Such compensators were initially proposed for controlling only the elastic motion [Ben.81, McL.87, Jos.89, Sla.90] of linear flexible space structures with no articulated appendages (i.e., single-body structures). These compensators were based on the fact that the plant consisting *only of elastic modes*, with velocity measurements as the output, is passive (i.e., the transfer function is positive-real(PR)), and can be stabilized by a strictly positive-real (in the strong sense) compensator. Even for the case of FDLTI systems certain problems occur with these results. First, the rigid-body attitude motion is not controlled, and second, the control law requires measurements of purely elastic motion, and assumes actuators that affect only the elastic motion. These assumptions do not hold for real spacecraft unless the actuators are used in a balanced configuration for accomplishing only damping enhancement with no rigid-body control [Jos.89].

Subsequently, the problem of controlling both rigid and elastic modes for linear, single-body flexible spacecraft was addressed in [Jos.95b], and was extended and generalized in [Jos.95] to multibody flexible spacecraft. Following Jos.95], the stability of dynamic dissipative compensators for flexible nonlinear multibody space structures in the attitude-hold configuration is proved next.

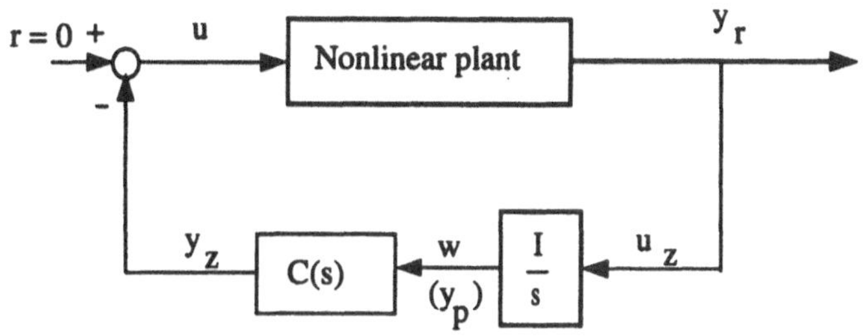

Figure 4.2: Feedback system: series connection

4.4.3.1 Stability Results

Consider the nonlinear system shown in Figure 4.2, which is described by Eqs. (4.33) and (4.35), controlled by an LTI controller with transfer function $C(s)$. The following theorem gives a sufficient condition for $C(s)$ to be a stabilizing controller. ($C(s)$ is said to stabilize the system $\dot{x} = f(x,u)$; $y = h(x,u)$, if the negative feedback interconnection of this system and a minimal realization of $C(s)$ is globally asymptotically stable).

Theorem 4.6 *The system given by Eqs. (4.33) and (4.35) with output y_p is stabilized by $C(s)$ if*

1) $C(s)$ is stable and minimum-phase

2) $[C(s)/s]$ is MSPR.

Proof 4.6 We shall first find a coordinate transformation which transforms the system (state vector) of Figure 4.2 to an equivalent system shown in Figure 4.3.

Because $C(s)$ is stable and $[C(s)/s]$ is MSPR, from Chapter 3, $[C(s)/s]$ can be decomposed as:

$$\frac{C(s)}{s} = \frac{C(0)}{s} + G_2(s) \tag{4.50}$$

where $C(0) = C(0)^T > 0$. Note that $C(0)$ is positive definite rather than semidefinite because $C(s)$ is minimum-phase. Because $[C(s)/s]$ is strictly proper,

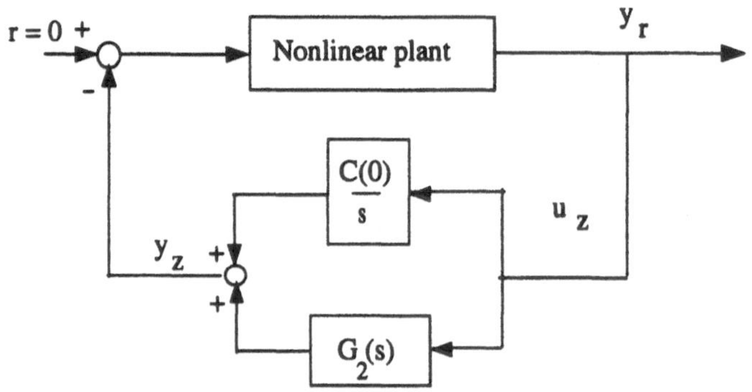

Figure 4.3: Feedback system: parallel connection

so is $G_2(s)$. Suppose $[\overline{A}_c, \overline{B}_c, \overline{C}_c, 0]$ is an n_c-th order minimal realization of $G_2(s)$. Let T denote an orthoganal matrix such that

$$T^T C(0) T = \Lambda \tag{4.51}$$

where Λ is a positive diagonal matrix consisting of the eigenvalues of $C(0)$. From Chapter 3, a minimal realization of $[C(s)/s]$ is:

$$\begin{bmatrix} \dot{\overline{w}} \\ \dot{\overline{x}}_c \end{bmatrix} := \dot{\overline{x}}_z = \begin{bmatrix} 0 & 0 \\ 0 & \overline{A}_c \end{bmatrix} \overline{x}_z + \begin{bmatrix} T^T \\ \overline{B}_c \end{bmatrix} u_z := \overline{A}_z \overline{x}_z + \overline{B}_z u_z \tag{4.52}$$

$$y_z = [T\Lambda \;\; \overline{C}_c] \overline{x}_z := \overline{C}_z \overline{x}_z \tag{4.53}$$

where u_z and y_z denote the $k \times 1$ input and output vectors. This realization corresponds to the "parallel" configuration (as shown in Figure 4.3) of $[C(s)/s]$.

Consider the transformation

$$\overline{x}_z = \begin{bmatrix} T^T & 0 \\ \overline{B}_c & I \end{bmatrix} x_z. \tag{4.54}$$

Then the above equations are transformed to:

$$\dot{x}_z := \begin{bmatrix} \dot{w} \\ \dot{x}_c \end{bmatrix} = \begin{bmatrix} 0 & 0 \\ B_c & A_c \end{bmatrix} x_z + \begin{bmatrix} I \\ 0 \end{bmatrix} u_z := A_z x_z + B_z u_z \tag{4.55}$$

$$y_z = [D_c \; C_c] x_z := C_z x_z \tag{4.56}$$

where $A_c = \overline{A}_c$, $B_c = \overline{A}_c \overline{B}_c$, $C_c = \overline{C}_c$, $D_c = C(0) + \overline{C}_c \overline{B}_c$. This representation corresponds to the "series" configuration: $[I/s].[C(s)]$, as shown in Figure 4.2, with $u_z = y_r$ and $w = y_p$.

From Chapter 3 Lemma 2, there exist matrices $\overline{P}_z = \overline{P}_z^T > 0$, $\overline{P}_z \in \Re^{(n_c+k)\times(n_c+k)}$, $L_c \in \Re^{k\times(n_c+k)}$ such that

$$\overline{A}_z^T \overline{P}_z + \overline{P}_z \overline{A}_z = -L^T L \tag{4.57}$$

$$\overline{C}_z = \overline{B}_z^T \overline{P}_z \tag{4.58}$$

$$L = [0_{k\times k}\ L_{c(k\times n_c)}] \tag{4.59}$$

where $[\overline{A}_c,\ L_c]$ is observable and $F(s) = L(sI - \overline{A}_z)^{-1}\overline{B}_z = L_c(sI - \overline{A}_c)^{-1}\overline{B}_c$ is minimum-phase.

The closed-loop system in Figure 4.2 is described by:

$$\frac{d}{dt}\begin{bmatrix} q \\ \dot{\gamma} \\ \dot{\theta} \\ \dot{q} \end{bmatrix} = \begin{bmatrix} 0 & 0\ \ 0\ \ I \\ -M^{-1}\begin{bmatrix} \tilde{K} \\ 0 \end{bmatrix} & -M^{-1}(C+D) \end{bmatrix} \begin{bmatrix} q \\ \dot{\gamma} \\ \dot{\theta} \\ \dot{q} \end{bmatrix} + \begin{bmatrix} 0 \\ M^{-1}B^T \end{bmatrix} u \tag{4.60}$$

$$\frac{d}{dt}\begin{bmatrix} \gamma \\ \theta \end{bmatrix} = \dot{y}_p = y_r = [0_{k\times(n-k)}\ I_k\ 0_{k\times(n-k)}]x \tag{4.61}$$

$$\dot{x}_c = A_c x_c + B_c \begin{bmatrix} \gamma \\ \theta \end{bmatrix} \tag{4.62}$$

$$y_z = C_c x_c + D_c \begin{bmatrix} \gamma \\ \theta \end{bmatrix} \tag{4.63}$$

$$u = -y_z. \tag{4.64}$$

This is a time-invariant system which has the form:

$$\dot{x} = f(x)$$

where $x = (q^T, \dot{\gamma}^T, \dot{\theta}^T, \dot{q}^T, y_p^T, x_c^T)^T = (q^T, \dot{p}^T, y_p^T, x_c^T)^T$. The dimension of the system is $(2n + n_c)$.

Apply the transformation

$$\overline{x} = \begin{bmatrix} I_{(2n-k)} & 0 & 0 \\ 0 & T_{k\times k}^T & 0 \\ 0 & \overline{B}_c & I_{n_c} \end{bmatrix} x \tag{4.65}$$

which transforms the state vector to: $\overline{x} = (q^T, \dot{p}^T, \overline{x}_z^T)^T$ where $\overline{x}_z = (\overline{w}^T, \overline{x}_c^T)$. Note that M and C are functions of $\overline{x}$.

Consider the candidate Lyapunov function

$$V(\overline{x}) = \frac{1}{2}\dot{p}^T M \dot{p} + \frac{1}{2} q^T \tilde{K} q + \frac{1}{2}\overline{x}_z^T P_z \overline{x}_z \tag{4.66}$$

where $\overline{P}_z$ is a positive definite matrix (Eq. 4.57). Differentiation with respect to t and simplification using Eq. (4.1), (4.57)-(4.59) leads to

$$\begin{aligned} \dot{V} &= -\dot{q}^T \tilde{D}\dot{q} + y_r^T u - \frac{1}{2}\overline{x}_z^T L^T L \overline{x}_z - y_r^T u \\ &= -\dot{q}^T \tilde{D}\dot{q} - \frac{1}{2}(L\overline{x}_z)^T (L\overline{x}_z) \\ &= -\dot{q}^T \tilde{D}\dot{q} - \frac{1}{2}(L_c\overline{x}_c)^T (L_c\overline{x}_c). \end{aligned} \tag{4.67}$$

Because $\tilde{D}$ is positive definite, $\dot{V} \leq 0$ (i.e., $\dot{V}$ is negative semidefinite) and the system is Lyapunov-stable. Now, $\dot{V} = 0$ only if $\dot{q} = 0$ and $L_c\overline{x}_c = 0$. Therefore, either $y_r = 0$ or y_r consists only of terms such as $\nu t^l e^{z_0 t}$ where ν is a constant vector and z_0 is a transmission zero of $(\overline{A}_c, \overline{B}_c, L_c)$. Because all transmission zeros of $(\overline{A}_c, \overline{B}_c, L_c)$ are in $[Re(s) < 0]$, $y_r \to 0$ exponentially. Since $(\overline{A}_c, \overline{B}_c, L_c)$ is minimal and stable, $\overline{x}_c \to 0$ exponentially, and $u = -y_z \to$ constant. $y_r \to 0$ also implies $\dot{\theta} \to 0$ and $\omega \to 0$, i.e., $\dot{p} \to 0$; this then implies that $\ddot{p} \to 0$. By substituting in equation (4.1), $q \to 0$. However, if $u \to$ nonzero constant, $\ddot{p}$ cannot tend to zero, i.e., $\dot{p}$ (and therefore y_r) would not tend to zero, which contradicts the fact that $y_r \to 0$; hence $u \to 0$, i.e., $\overline{w} \to 0$, and $\overline{x}_z \to 0$. Therefore, along the trajectories for which $\dot{V} = 0$, we have $\dot{p} \to 0$, $q \to 0$, $\overline{x}_z \to 0$, and $\overline{w} \to 0$, i.e., $\overline{x} \to 0$, and V *decreases* (and $\dot{V} < 0$) along some part of the trajectory. Therefore such trejectories (other than $\overline{x} \equiv 0$) do not exist, and by LaSalle's invariance principle the system is globally asymptotically stable. □

Because no assumptions were made with regard to the model order as well as to the knowledge of the parametric values, the stability is robust to modeling errors and parametric uncertainties.

Remark: The controller $C(s)$ stabilizes the complete plant; i.e., the system consisting of the rigid modes, the elastic modes, and the compensator state vector (x_c) is g.a.s. The global asymptotic stability is guaranteed regardless of the number of modes in the model or parameter uncertainties, i.e., the stability is *robust* to modeling errors. The order of the controller can be chosen to be any number $\geq k$. In other words, these results enable the design of a controller of essentially any desired order, which robustly stabilizes the plant. (Note that it would also be possible to prove Theorem 4.6 by applying Theorem 3.8 for the feedback system consisting of (4.60), and (4.61-4.64). However, it would be necessary to first show that the system (4.60) with output y_r is internally passive and strongly zero-state observable). The condition that $[C(s)/s]$ be MSPR is fairly straightforward to check if $C(s)$ is diagonal. Let $C(s) = \text{diag}[C_1(s), \ldots, C_k(s)]$, where

$$C_i(s) = k_i \frac{s^2 + \beta_{1i}s + \beta_{0i}}{s^2 + \alpha_{1i}s + \alpha_{0i}}. \tag{4.68}$$

A straightforward analysis shows that $[C(s)/s]$ is MSPR if, and only if, k_i, α_{0i}, α_{1i}, β_{0i}, and β_{1i} are positive for $i = 1, 2, \ldots, k$, and

$$\alpha_{1i} - \beta_{1i} > 0 \tag{4.69}$$

$$\alpha_{1i}\beta_{0i} - \alpha_{0i}\beta_{1i} > 0. \tag{4.70}$$

For higher order C_i's, the conditions on the polynomial coefficients are harder to obtain. One systematic procedure for obtaining such conditions for higher order controllers is the application of Sturm's theorem [Van.65]. Symbolic manipulation codes can then be used to derive explicit inequalities. The controller design problem can be subsequently posed as a constrained optimization problem which minimizes a given performance function. However, the case of fully populated $C(s)$ has no straightforward method of solution and remains an area for future research.

The following results, which address the cases with static dissipative controllers when the actuators have first- and second-order dynamics, are an immediate consequence of Theorem 4.6 and are stated without proof.

Corollary 5.1: For the static dissipative controller (Eq. (4.34)), suppose that $\overline{G}_p$ and G_r are diagonal with positive entries denoted by subscript i, and actuators represented by the transfer function $G_{Ai}(s) = k_i/(s + a_i)$ are present in the ith control channel. Then the closed-loop system is g.a.s. if $G_{ri} > \overline{G}_{pi}/a_i$ (for $i = 1, 2, \ldots, k$).

Corollary 5.2: Suppose that the static dissipative controller also includes the feedback of the acceleration y_a, that is,

$$u = -\overline{G}_p y_p - G_r y_r - G_a y_a \tag{4.71}$$

where $\overline{G}_p$, G_r, and G_a are diagonal with positive entries. Suppose that the actuator dynamics for the ith input channel are given by $G_{Ai}(s) = k_i/(s^2 + \mu_i s + \nu_i)$ with k_i, μ_i, and ν_i positive. Then the closed-loop system is a.s. if

$$\frac{G_{ri}}{G_{ai}} \le \mu_i < \frac{G_{ri}}{\overline{G}_{pi}} \quad (i = 1, 2, \ldots, k).$$

4.4.3.2 Realization of $C(s)$ as strictly proper controller

The controller $C(s)$ (eqs. (4.62) and (4.63)) is not strictly proper because of the direct transmission term D_c. From a practical viewpoint, a strictly proper controller is sometimes desirable because it attenuates sensor noise as well as high-frequency disturbances. Furthermore, the most common types of controllers, which include the linear quadratic Gaussian (LQG) controllers as well as the observer-pole placement controllers, are strictly proper; they have a first-order rolloff. Direct implementation of $C(s)$ does not utilize the rate measurement y_r. The following result states that $C(s)$ can be realized as a strictly proper controller wherein both y_p and y_r are utilized.

Theorem 4.7 *The nonlinear plant with y_p and y_r as outputs is stabilized by the controller C' given by*

$$\dot{\hat{x}}_c = A_c \hat{x}_c + [\, B_c - A_c Q \quad Q \,] \begin{bmatrix} y_p \\ y_r \end{bmatrix} \tag{4.72}$$

$$y_c = C_c \hat{x}_c \tag{4.73}$$

where C_c is assumed to be of full rank, and an $n_c \times k$ ($n_c \ge k$) matrix Q is a solution of

$$D_c - C_c Q = 0. \tag{4.74}$$

Proof 4.7 Consider the controller realization equations (4.62) and (4.63). Let

$$\hat{x}_c = x_c + Qy_p \tag{4.75}$$

where Q is an $n_c \times k$ matrix. The differentiation of equation (4.75), use of equations (4.62) and (4.63), and replacement of $\dot{y}_p$ with y_r results in equation (4.72) and leads to

$$y_c = C_c\hat{x}_c + (D_c - C_cQ)\, y_p. \tag{4.76}$$

If Q is chosen to satisfy equation (4.74), the strictly proper controller is given by equations (4.72) and (4.73). Equation (4.74) represents k^2 equations in kn_c unknowns. Because $k < n_c$ (i.e., the compensator order is greater than the number of plant inputs) and C_c is of full rank, many possible solutions exist for Q. The solution which minimizes the Frobenius norm of Q is

$$Q = C_c^T \left(C_cC_c^T\right)^{-1} D_c. \tag{4.77}$$

If $k = n_c$, equation (4.77) gives the unique solution $Q = C_c^{-1}D_c$. □

4.5 Linear Single-Body Spacecraft

In this section, we consider the case of a single-body flexible spacecraft, which is a special case of the systems discussed in the previous section. The motivation for investigating this case separately is that a large number of important spacecraft configurations belong to this class (e.g., flexible space antennas). In addition, the mathematical models for this class of spacecraft are linear, which permits the use of a variety of controller synthesis techniques that are not available for nonlinear systems. However, the stability results for the linear case under static and dynamic dissipative compensation can be obtained as special cases of the stability results for nonlinear models.

4.5.1 Linearized Mathematical Model

The linearized mathematical model of the rotational dynamics of single-body flexible space structure can be obtained by linearizing equation (4.1) about the zero steady state and is given by

$$\widehat{M}\ddot{p} + \widehat{K}p = B^T u \tag{4.78}$$

where

$$p = \left(\eta^T, q^T\right)^T \qquad \eta = (\phi, \theta, \psi)^T . \tag{4.79}$$

Here, η represents the 3×1 attitude vector, q is the $n_q \times 1$ vector of elastic modal amplitudes ($n_q = n - 3$), $\widehat{M}$ is the positive definite symmetric mass-inertia matrix (note that $\widehat{M}$ is constant in this case), $\widehat{K}$ is the positive semidefinite stiffness matrix related to the flexible degrees of freedom, u is the 3×1 input torque vector, and $B = [I_3, 0_{3\times n_q}]$. The system in equation (4.78) can be transformed into modal form by using transformation $\zeta = \Phi p$, where Φ is the eigenvector matrix such that $\Phi^T \widehat{M} \Phi = I$ and $\phi^T \widehat{K} \Phi = \widetilde{C}$, a diagonal matrix. The resulting model is

$$\ddot{\zeta} + \widetilde{C}\zeta = \Phi^T B^T u = \begin{bmatrix} \overline{\Phi}_{11}^T \\ \overline{\Phi}_{12}^T \end{bmatrix} u \tag{4.80}$$

where $\overline{\Phi}_{11}^T$ and $\overline{\Phi}_{12}^T$ are 3×3 and $n_q \times 3$ matrices, respectively, and matrix $\widetilde{C}$ is given by

$$\widetilde{C} = \operatorname{diag}(0_3, \Lambda) \tag{4.81}$$

where

$$\Lambda = \operatorname{diag}\left(\omega_1^2, \omega_2^2, \ldots, \omega_{n_q}^2\right) \tag{4.82}$$

in which ω_i $(i = 1, 2, \ldots, n_q)$ represent the elastic mode frequencies. The first three components of ζ correspond to rigid-body modes. The rigid-body modes are controllable if, and only if, $\overline{\Phi}_{11}$ is nonsingular. Because one torque actuator is used for each of the X-, Y-, and Z-axes, $\overline{\Phi}_{11}$ is nonsingular. Note that the mass-inertia matrix in this modal form is the 3×3 identity matrix. However, the model is customarily used in a slightly modified form wherein the elastic motion is superimposed on the rigid-body motion. (See [Jos.89].) This can be achieved as follows. Suppose $J = J^T > 0$ is the moment-of-inertia matrix of the spacecraft. Then, with the use of the transformation, $\zeta = \Delta\xi$ where the transformation matrix Δ is given by

$$\Delta = \begin{bmatrix} \overline{\Phi}_{11}^T J & 0_{3\times n_q} \\ 0_{n_q\times 3} & I_{n_q\times n_q} \end{bmatrix}. \tag{4.83}$$

Equation (4.80) is transformed to

$$\Delta\ddot{\xi} + \widetilde{C}\Delta\xi = \begin{bmatrix} \overline{\Phi}_{11}^T \\ \overline{\Phi}_{12}^T \end{bmatrix} u. \tag{4.84}$$

Premultiplication of the above equation by Δ^{-1} yields

$$\ddot{\xi} + \Delta^{-1}\widetilde{C}\Delta\xi = \begin{bmatrix} J^{-1} \\ \overline{\Phi}_{12}^T \end{bmatrix} u. \tag{4.85}$$

Note that $\Delta^{-1}\widetilde{C}\Delta = \widetilde{C}$ and with premultiplication by $\text{diag}(J, I_{n_q})$, equation (4.85) can be rewritten as

$$\widetilde{A}\ddot{\xi} + \widetilde{C}\xi = \Gamma^T u \tag{4.86}$$

where $\widetilde{A} = \text{diag}(J, I)$ and $\Gamma = [I_3, \overline{\Phi}_{12}]$. The inherent structural damping term $\widetilde{B}$ can now be added to give the design model as follows:

$$\widetilde{A}\ddot{\xi} + \widetilde{B}\dot{\xi} + \widetilde{C}\xi = \Gamma^T u \tag{4.87}$$

where

$$\widetilde{A} = \text{diag}\left(J, I_{n_q}\right) \tag{4.88}$$

$$\widetilde{B} = \text{diag}\left(0_3, D\right) \tag{4.89}$$

$$\widetilde{C} = \text{diag}\left(0_3, \Lambda\right) \tag{4.90}$$

$$\Lambda = \text{diag}\left(\omega_1^2, \omega_2^2, \ldots, \omega_{n_q}^2\right). \tag{4.91}$$

$D = D^T > 0$ is the $n_q \times n_q$ matrix representing the inherent damping in the elastic modes, ω_i $(i = 1, 2, \ldots, n_q)$ represent the elastic mode frequencies, and

$$\Gamma = [\, I_{3\times 3} \quad \overline{\Phi}_{12} \,]. \tag{4.92}$$

The attitude and attitude rate sensor outputs are given by

$$y_p = \Gamma\xi \tag{4.93}$$

$$y_r = \Gamma\dot{\xi} \tag{4.94}$$

where y_p and y_r are 3×1 measured position and rate vectors, respectively.

All of the stability results of sections 4.4 for static and dynamic dissipative controllers are directly applicable to this case. From Theorem 4.4, the constant gain dissipative control law is given by

$$u = -G_p y_p - G_r y_r \tag{4.95}$$

where G_p and G_r, which are symmetric, positive definite, proportional and rate gain matrices, respectively, make the closed-loop system asymptotically stable. Furthermore from Theorem 4.5, the stability is robust in the presence of monotonically nondecreasing actuator nonlinearities and sensor nonlinearities belonging to the $(0, \infty)$ sector.

For this case, note that the transfer function is given by $G(s) = G'(s)/s$ (Figure 4.4), where $G'(s)$ is given by

$$G'(s) = \frac{J^{-1}}{s} + \sum \frac{\Phi_i \Phi_i^T s}{s^2 + 2\rho_i \omega_i s + \omega_i^2} \tag{4.96}$$

where J is the moment-of-inertia matrix, and Φ_i, ρ_i, and ω_i denote the rotational mode shape matrix, damping ratio, and natural frequency, respectively, of the ith structural mode. The operator represented by the transfer function $G'(s)$ is passive. Because this operator is linear and time-invariant, its passivity implies that the transfer function $G'(s)$ is positive-real. However, the transfer function $G(s)$, from u to y_p, is not positive-real. Assuming $\rho_i > 0$, Theorem 4.6 can be applied to show that the system is robustly stabilized by dynamic dissipative compensator $C(s)$ if $C(s)$ is stable and $[C(s)/s]$ is MSPR. However, Theorem 3.4 can be readily applied to show robust stability of the system even when $\rho_i \geq 0$. $[C(s)/s]$ has $j\omega$ -axis poles only at $s = 0$, and $G'(s)$ does not have transmission zeros at $s = 0$, therefore from Theorem 3.4, $[C(s)/s]$ stabilizes G'. This implies that $C(s)$ stabilizes $G(s)$. The compensator can be realized as a strictly proper one as shown in Theorem 4.7.

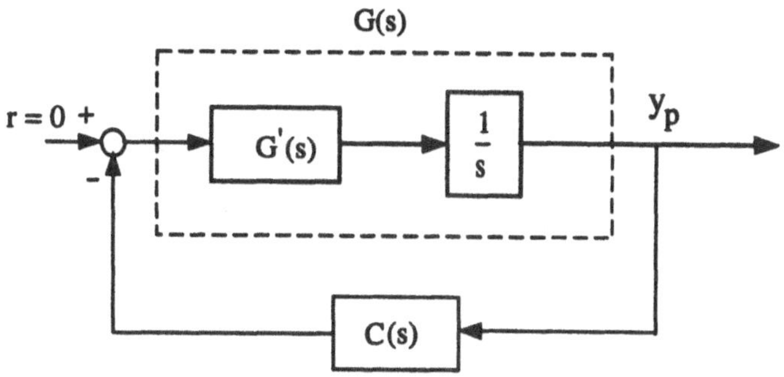

Figure 4.4: Feedback loop for linear case

4.5.2 Optimal Dynamic Dissipative Compensator

The results of Theorems 4.6 and 4.7 can be applied to check if a given LQG controller is dissipative (i.e., if it robustly stabilizes $G(s)$). In particular, the following result is obtained.

Theorem 4.8 *Consider the n_cth-order LQG controller given by*

$$\dot{\varsigma} = A_\varsigma \varsigma + H \begin{bmatrix} y_p \\ y_r \end{bmatrix} \tag{4.97}$$

$$u = F\varsigma \tag{4.98}$$

where A_ς is the closed-loop LQG compensator matrix

$$A_\varsigma = A_0 - B_0 F + H C_0. \tag{4.99}$$

A_0, B_0, and C_0 denote the design model matrices, and F and H are the regulator and estimator gain matrices, respectively. This controller robustly stabilizes the system if the rational matrix $M(s)/s$ is MSPR where

$$M(s) = F(sI - A_\varsigma)^{-1}(H_1 + A_\varsigma H_2) + FH_2 \tag{4.100}$$

and H_1 and H_2 denote the matrices consisting of the first three columns and the last three columns of H, respectively.

The theorem can be proved by using the transformation $\dot{\tilde{x}} = \varsigma - H_2 y_p$ in equation (4.97). Although given LQG controller will not likely satisfy the condition of Theorem 4.8, the condition can be incorporated as a constraint in the design process. The problem can be posed as one of minimizing a given LQG performance function with the constraint that $[M(s)/s]$ is MSPR. Also note that Theorem 4.8 is not limited to an LQG controller but is valid for any observer-based controller with control gain F and observer gain H. Another way of posing the design problem is to obtain the dissipative compensator which is closest to a given LQG design. The distance between compensators can be defined as either

1. The distance between the compensator transfer functions in terms of H_2 or H_∞ norm of the difference or
2. The distance between the matrices used in the realization in terms of a matrix (i.e., spectral or Frobenius) norm

For example, the dissipative compensator A_c and C_c matrices can be taken to be the LQG A_ς and F matrices, respectively, and the compensator B_c and D_c matrices can be chosen to minimize δ as follows while still satisfying the MSPR constraint:

$$\delta = \left\| \begin{matrix} B_c - (H_1 + A_\varsigma H_2) \\ D_c - FH_2 \end{matrix} \right\|_{2 \text{ or } \infty} . \tag{4.101}$$

Thus the design method usually ends up as a constrained optimization problem.

The next section presents numerical examples for demonstrating the various nonlinear and linear control laws discussed.

4.6 Numerical Examples

Two numerical examples are given to demonstrate some of the results obtained in sections 4.3 and 4.5. The first example consists of a conceptual nonlinear model of a spacecraft with two flexible articulated appendages. The stability results obtained for nonlinear dissipative control laws given in section 4.3 are verified by numerical simulation. The simulation results were obtained using both the control laws (Eqs. 4.16 and 4.25), one of which uses a fully populated gain matrix (Eq. 4.17) while the other uses only a scalar gain (Eq. 4.26). It is shown that the control law with fully populated gain matrix gives better performance than the scalar gain control law, as would be expected. The reason is that, the fully populated gain matrix allows the control designer additional degrees of freedom which can be used to enhance the performance.

The second example addresses attitude control system design for a large space antenna, which is modeled as a linear single-body structure. The objective of the control system is to minimize a prescribed quadratic performance index. For this system, the conventional LQG controller design was found to have stability problems due to unmodeled high-frequency elastic modes and parametric uncertainties. However, the dynamic dissipative controller designed to minimize the quadratic performance function resulted in satisfactory performance with guaranteed stability in the presence of both unmodeled dynamics and parametric uncertainties.

4.6.0.1 Two-Link Flexible Space Robot

The system shown in Figure 4.5 is used for validation of some of the theoretical results obtained in section 4.3. The configuration consists of a central body with two articulated flexible links attached to it, and resembles a flexible space robot. The central body is a solid cyinder 1.0 m in diameter and 2.0 m in height. Each link is modeled in MSC/NASTRAN [Ano.92] as a 3.0 m-long flexible beam with 20 bar elements. The circular cross sections of the links are 1.0 cm in diameter resulting in significant flexibility. The material chosen for the central body as well as the links has a mass density of 2.568×10^{-3} kg/m^3 and modulus of elasticity $E = 6.34 \times 10^9$ kg/m^2. The central-body mass is 4030 kg and each link mass is 0.605 kg. The principal moments of inertia of the central-body about the local X-, Y-, and Z-axes are 1600, 1600, and 500 kg-m2, respectively. Each link can can rotate about its local Z-axis; the link moment of inertia about its axis of rotation is 1.815 kg-m^2. The central body has three rotational degrees of freedom. As shown in Figure 4.5, there are two revolute joints: one between the central-body and link 1 and another between links 1 and 2. The axes of rotation for revolute joints 1 and 2 coincide with the local Z-axes of links 1 and 2, respectively. Collocated torque actuators and

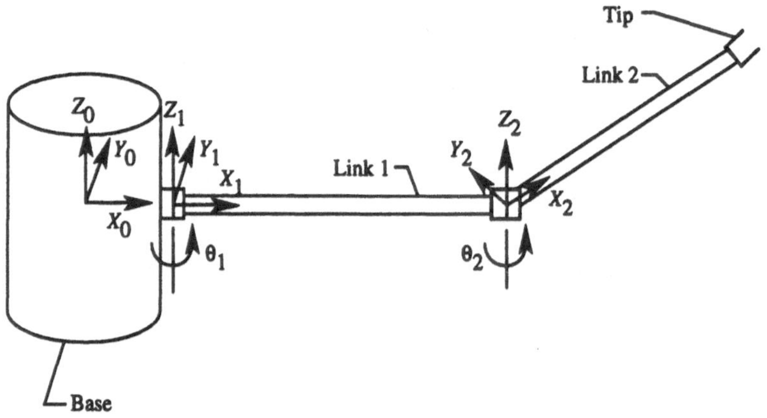

Figure 4.5: Flexible space robot

rotational sensors are assumed for each rigid degree of freedom. The sensor measurements consist of the central-body attitude (quaternions) and rates, as well as the revolute joint angles and rate.

The first link was modeled as a flexible beam with pinned-pinned boundary conditions; the second link was modeled as a flexible beam with pinned-free boundary conditions. For the purpose of simulation, the first four bending modes in the local XY-plane were considered for each link (i.e., the system has five rigid rotational degrees of freedom and eight flexible degrees of freedom, four for each link). The modal data were obtained from MSC/NASTRAN [Ano.92]. The mode shapes at the frequencies noted for links 1 and 2 are shown in Figures 4.6 and 4.7, respectively. A complete nonlinear simulation was obtained with DADS [Ano.89].

A rest-to-rest maneuver was considered to demonstrate the control laws. The initial configuration was equivalent to $(\pi/4)$-rad rotation of the entire spacecraft about the global X-axis and 0.5-rad rotation of the revolute joint 2. The objective of the control law was to restore the zero state of the system.

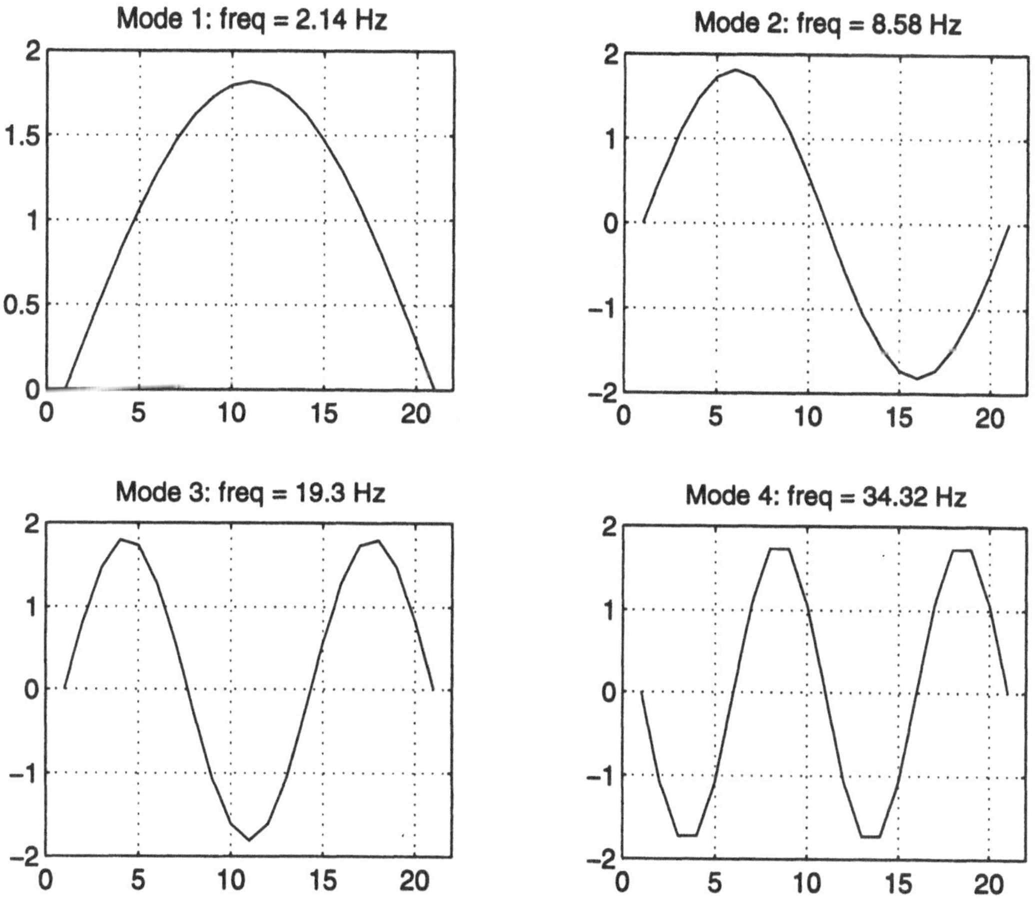

Figure 4.6: Mode shapes of link-1

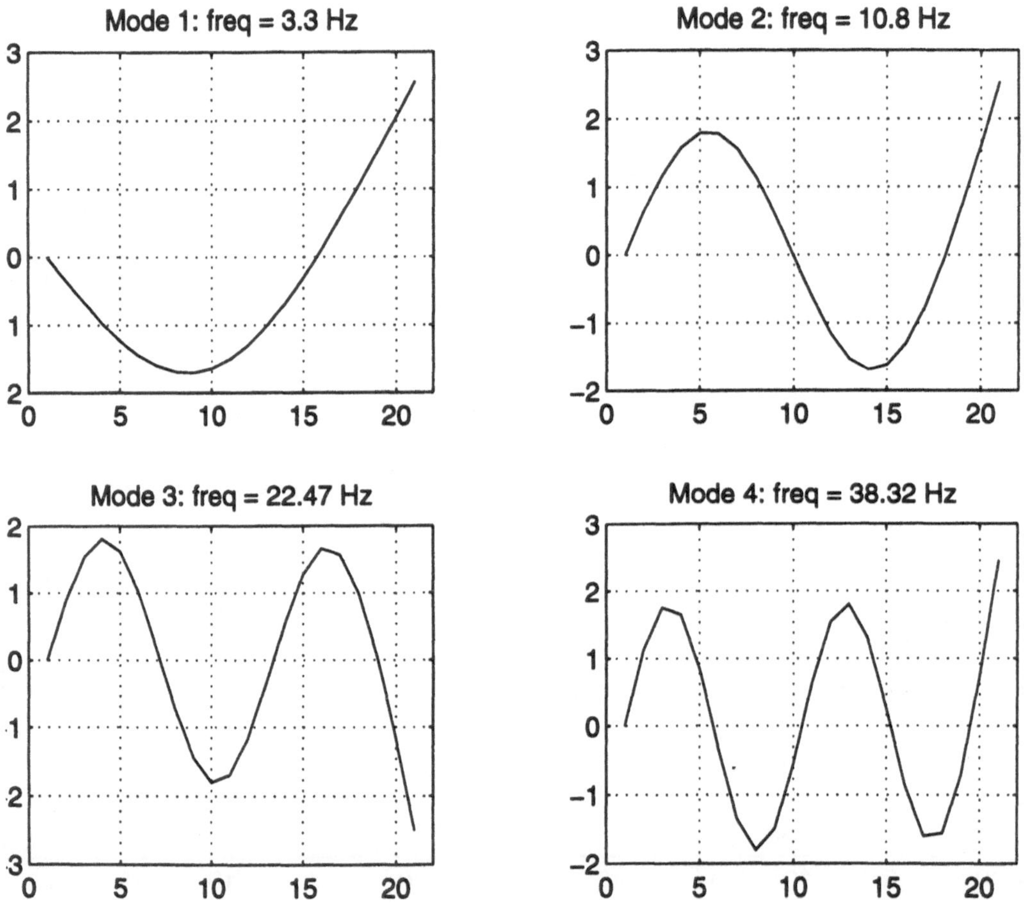

Figure 4.7: Mode shapes of link-2

Nonlinear dissipative controllers given by Eqs. (4.16) and (4.25) were used to accomplish the task. Because there are no known techniques to date for the synthesis of such controllers, the selection of controller gains was based on trial and error. Based on several trials, for the control law of Eq. (4.16), the following gains were found to give the desirable response:$G_{p1} = \text{diag}[550, 425, 350]$, $G_{p2} = \text{diag}[60, 70]$, $G_r = \text{diag}[450, 275, 300, 80, 90]$, and $\nu = 1000$. Similarly, for the control law of Eq. (4.25), the suitable set of gains were found to be: $G_{p1} = \text{diag}[1000, 1000, 1000]$, $G_{p2} = \text{diag}[50, 50]$, $G_r = \text{diag}[500, 275, 270, 100, 100]$. As the system begins motion, all the members move relative to one another, and dynamic interaction exists between the members. Complete nonlinear and coupling effects are incorporated in the simulations. Figures 4.8-4.10 show the responses for Euler parameter 1, displacement of revolute joint 2, and torque for Euler axis 1 for both control laws. Remaining reponses are shown only for the control law of Eq. (4.16) since there were no noticable differences in the two reponses. The Euler parameter responses for Euler axes 2 and 3 are shown in Figures 4.11 and 4.12. The joint angle displacement for the revolute joint 1 is shown in Figure 4.13. The joint displacements decay asymptotically and are nearly zero within 15 sec. The tip displacements with respect to global X-, Y-, and Z-axes are shown in Figures 4.14 and 4.15. Note that the manipulator tip reaches its desired X position in about 15 sec, whereas the desired Y and Z positions are reached in about 35 sec. The time histories of control torques are given in Figures 4.16-4.17. The effects of nonlinearities in the model can be seen in the responses as well as in the torque profiles.

Note that, although the position gains used in control law (4.16) for the central body attitude (G_{p1}) are almost half of those used in (4.25), the response has improved, resulting in less overshoot for the same settling time. This improvement can be attributed to the additional terms and the additional design freedom available in the choice of G_{p1} and ν with the control law of Eq. (4.16). As a consequence, the control torque for Euler axis 1 has significantly reduced as shown in Figure 4.10. Similar improvement was observed in the displacement response of revolute joint 2. With the control law of Eq. (4.16),

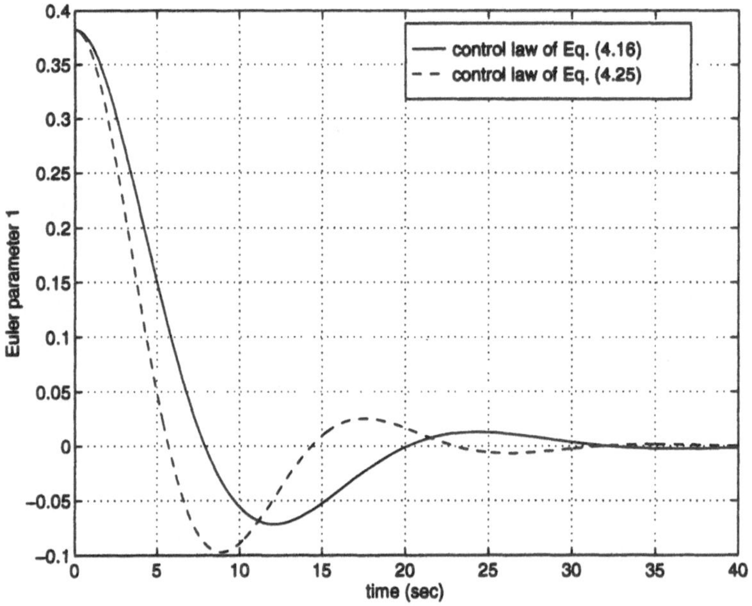

Figure 4.8: Euler parameter 1

the overall saving achieved in the control energy ($= \int_0^{40} u^T u \; dt$) was 40% as compared to the control law of Eq. (4.25), along with significant improvement in the responses. The response for the $y-$displacement of the link 2-tip showed slightly more overshoot and oscillations; however, the settling time was not greatly affected.

In spite of the performance differences, both the control laws effectively demonstrate the stability results of section 4.3.

4.6.0.2 Application to Hoop-Column Antenna

The 122-m-diameter, hoop-column antenna concept (Figure 1.1), as described in [Jos.89], consists of a deployable mast attached to a deployable hoop by tension cables. The antenna has many significant elastic modes, which include mast bending, torsion, and reflective surface distortion. The objective is to control the attitude (including rigid and elastic components) at a certain point on the mast in the presence of actuator noise and attitude and rate sensor noise.

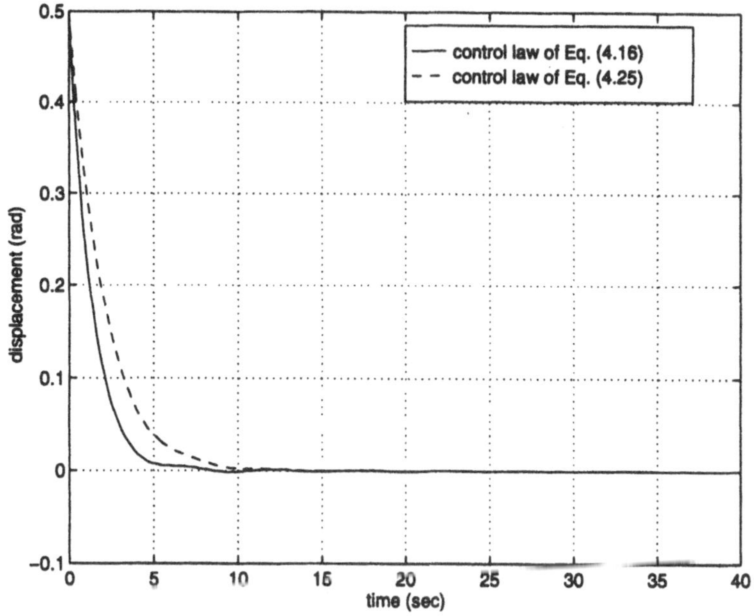

Figure 4.9: Revolute joint-2 displacement

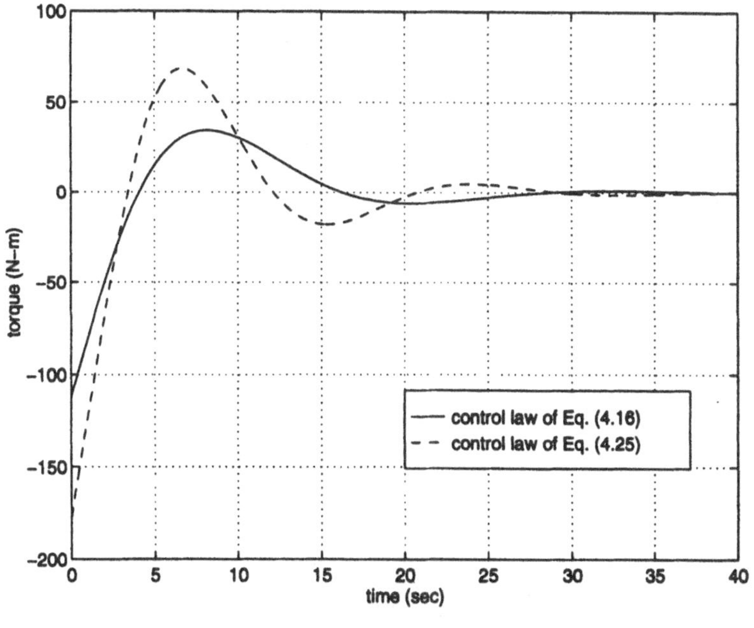

Figure 4.10: Torque for Euler axis 1

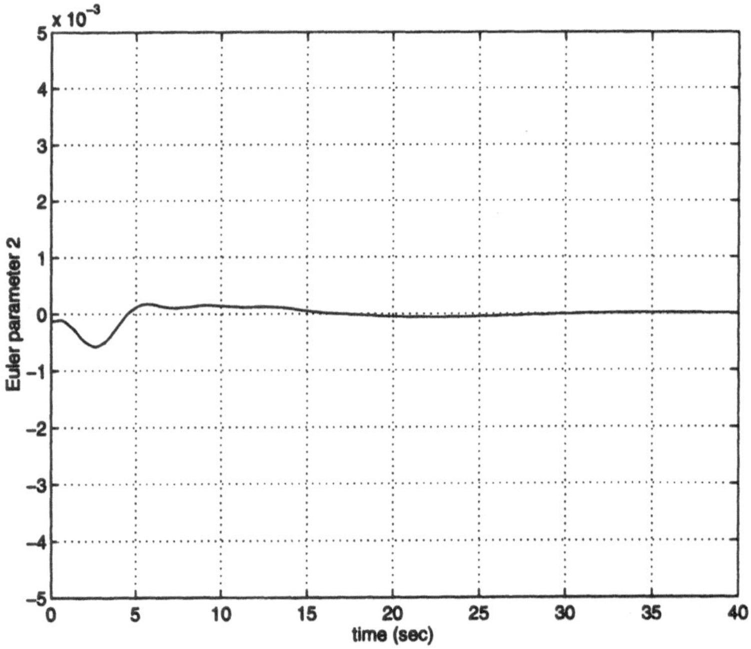

Figure 4.11: Euler parameter 2

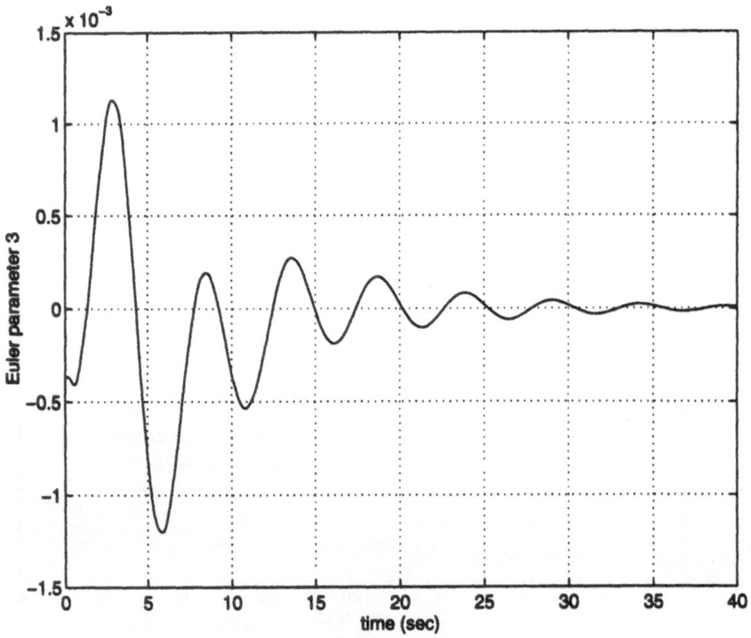

Figure 4.12: Euler parameter 3

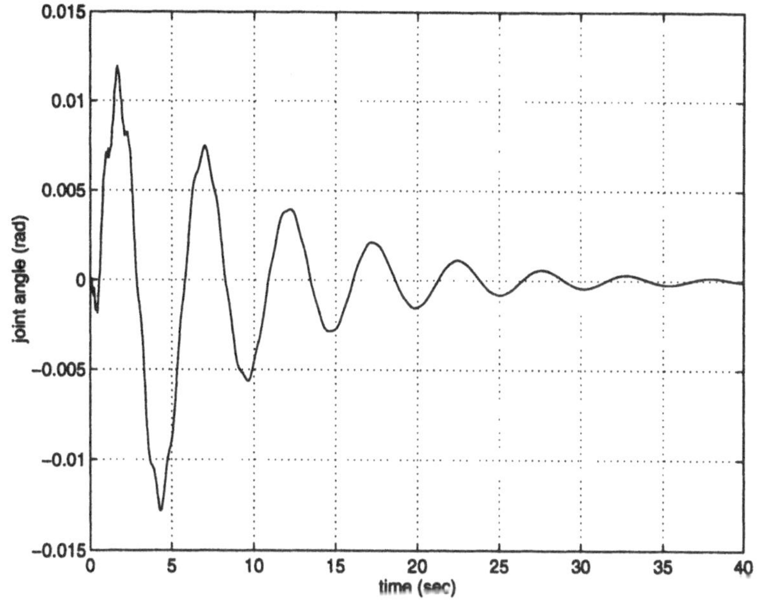

Figure 4.13: Revolute joint-1 displacement

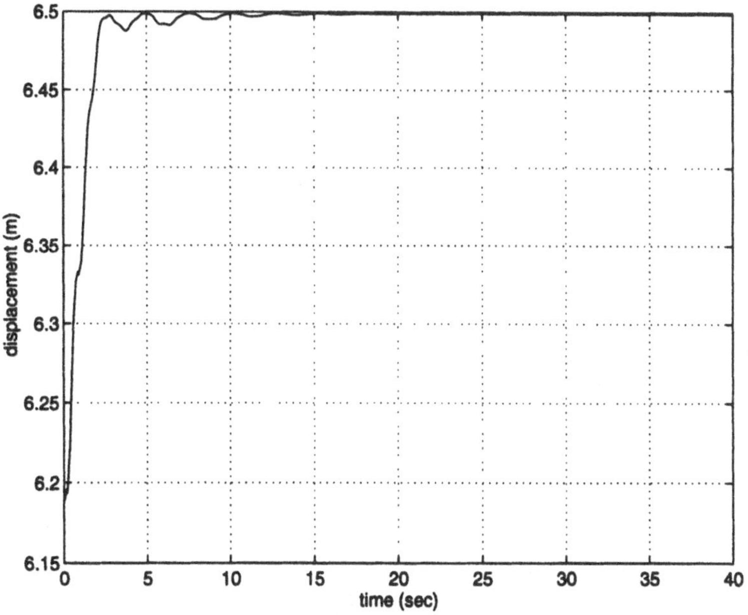

Figure 4.14: x-displacement of tip

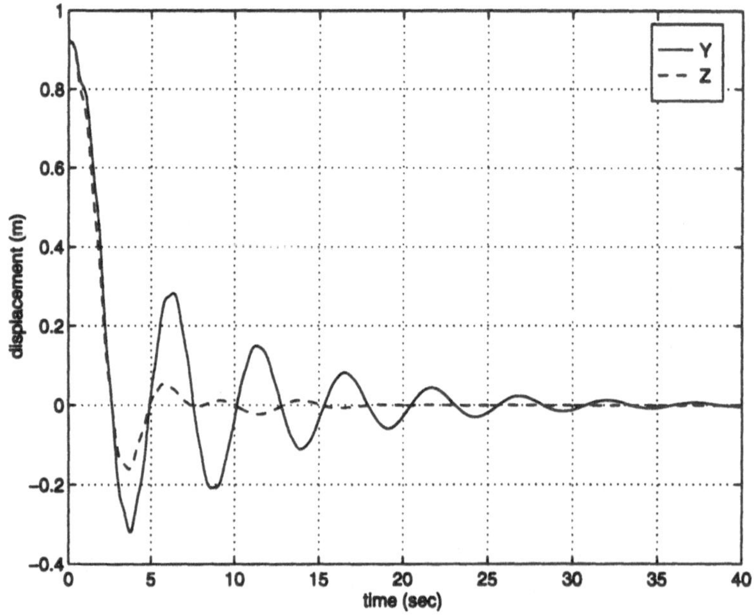

Figure 4.15: y and z-displacements of tip

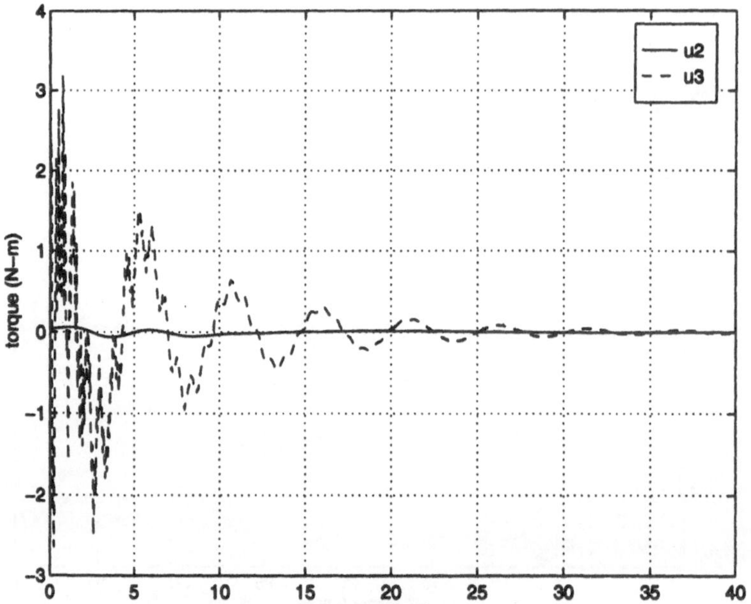

Figure 4.16: Torques for Euler axes 2 & 3

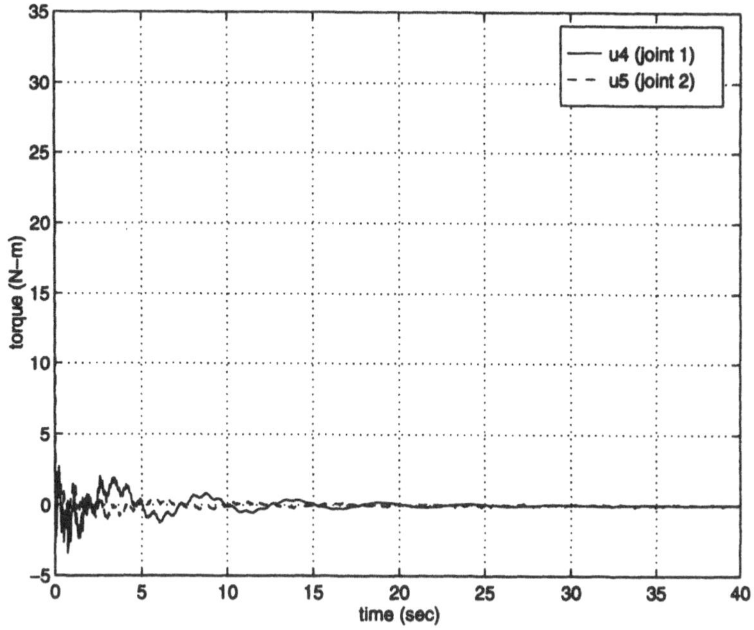

Figure 4.17: Torques at revolute joints

One attitude sensor and one rate sensor is collocated with a torque actuator on each of the three axes. The open-loop damping ratio is assumed to be 1 per cent. A linear quadratic Gaussian (LQG) controller, based on a design model consisting of the three rotational rigid modes and the first three elastic modes, was first designed to minimize

$$\mathcal{J} = \lim_{T\to\infty} \frac{1}{T}\mathcal{E} \int_0^T \left(y_p^T Q_p y_p + y_r^T Q_r y_r + u^T R u\right) dt \tag{4.102}$$

with $Q_p = 4 \times 10^8 I_3$, $Q_r = 10^8 I_3$, and $R = \text{diag}(0.1, 0.1, 1)$. The actuator noise covariance intensity was $0.1 I_3$ (ft-lb)^2 and the attitude and rate sensor noise covariance intensity was $10^{-10}\text{diag}(0.25, 0.25, 2.5)$ rad/sec and $10^{-10}\text{diag}(0.25, 0.25, 2.5)$ rad^2/sec^2, respectively. The optimal value of $\mathcal{J}$ was 0.6036, and the closed-loop eigenvalues for the design model and the 12th-order controller are given in Table 4.1.

A dynamic dissipative controller, which consisted of three second-order blocks as in equation (4.68), was designed next. By using the transformation

Table 4.1: Closed-Loop Eigenvalues for LQG Controller

Regulator	Estimator
-0.0238±0.0542i	-0.0240±0.0544i
-0.0721±0.0964i	-0.0720±0.0965i
-0.0754±0.1000i	-0.0725±0.0963i
-0.3169±0.8108i	-0.4050±0.7703i
-0.2388±1.3554i	-0.3334±1.3333i
-0.3375±1.7030i	-0.5104±1.6562i

of Theorem 4.7 with $L_i = [\gamma_i, \delta_i]^T$ for $C_i(s)$, each $C_i(s)$ can be realized as a strictly proper controller

$$\dot{\overline{x}}_{ci} = \begin{bmatrix} 0 & 1 \\ -\alpha_{0i} & -\alpha_{1i} \end{bmatrix} \overline{x}_{ci} + \begin{bmatrix} \delta_i & \gamma_i \\ \beta_{0i}\gamma_i + \beta_{1i}\delta_i & \delta_i \end{bmatrix} \begin{bmatrix} y_{pi} + w_{pi} \\ y_{ri} + w_{ri} \end{bmatrix} \tag{4.103}$$

$$u = (u_1, u_2, u_3)^T \qquad u_i = (\beta_{0i} - \alpha_{0i})\,\overline{x}_{ci} \tag{4.104}$$

where w_{pi} and w_{ri} denote zero-mean white noise in the position and rate sensors. The constraints to be satisfied are equation (4.69), equation (4.70) and that α_{0i}, α_{1i}, β_{0i}, and β_{1i} be positive ($i = 1, 2, 3$). Thus, this sixth-order compensator has 18 design variables. The performance function in equation (4.102) can be computed by solving the steady-state covariance equation for the closed-loop state equation for the plant and the controller. A dynamic dissipative controller (DDC) was designed by performing numerical minimization of the performance function $\mathcal{J}$ with respect to the 18 design variables. To ensure a reasonable transient response, an additional constraint that the real parts of the closed-loop eigenvalues be ≤ -0.0035 is imposed. Table 4.2 lists the resulting closed-loop eigenvalues. Although the value of $\mathcal{J}$ for the DDC was 1.2674 (about twice that for the LQG controller), the closed-loop eigenvalues indicate satisfactory damping ratios and decay rates. Furthermore, the LQG controller, which was based on the first six modes, caused instability when higher modes were included in the evaluation model, whereas the DDC yields guaranteed

Table 4.2: Closed-Loop Eigenvalues for DDC

-0.0035±0.0194i
-0.0183±0.0458i
-0.0160±0.0502i
-0.3419±0.5913i
-0.7179±0.6428i
-0.8479±0.5653i
-0.6482±1.6451i
-0.4536±2.1473i
-0.3764±2.5522i

stability in the presence of higher modes as well as parametric uncertainties.

4.7 Summary and Remarks

This chapter addressed the problem of controlling the class of nonlinear multi-body flexible space systems described in Chapter 2. The system consists of a flexible central body to which a number of articulated appendages are attached. Assuming collocated torque actuators and angular position and rate sensors, global asymptotic stability of such systems is established using nonlinear dissipative control laws which employ the feedback of the central-body quaternion and the measured angles and rates. The stability proofs use the Lyapunov approach and exploit the inherent passivity of such systems. The stability is shown to be robust to unmodeled dynamics and parametric uncertainties. For a special case in which the attitude motion of the central body is small, the system, although still nonlinear, can be robustly stabilized by *linear* static and dynamic dissipative controllers. The static dissipative control law is shown to provide robust stability in the presence of certain classes of actuator and sensor nonlinearities and actuator dynamics. The results obtained for this special case are shown to represent a generalization of the results established in [Jos.89] for single-body linear flexible space structures. For this case, a synthesis technique for the design of a suboptimal dynamic dissipative controller is also presented. All the stability results presented are valid in spite of unmodeled modes and parametric uncertainties, i.e., the stability is robust to model errors and uncertainties. The results have a significant practical value because the mathematical models of such systems usually have substantial inaccuracies, and the actuation and sensing devices have nonlinearities. The results obtained are applicable to a broad class of multibody and single-body systems such as flexible multilink manipulators, multipayload space platforms, and space antennas.

Chapter 5

Trajectory Tracking of Spacecraft

5.1 Introduction

In the previous chapter, we presented various stabilizing control laws for single body and multibody flexible spacecraft. The problem addressed was of bringing the spacecraft from any initial state to a desired final resting position. In many applications, however, it is necessary to follow a certain desired trajectory during a spacecraft maneuver. For example, manipulators aboard a space platform may need to track some predetermined trajectory, or some moving payloads aboard research satellites may be required to follow certain prescribed path during a maneuver. Such applications motivate the need for stable tracking control laws for spacecraft maneuvers, which is the subject of this chapter.

The problem of trajectory control of multibody systems has been addressed widely in the robotics literature. In particular, there has been substantial work done in the area of tracking control of terrestrial robots [Sin.85, Pad.90, Wen.88, Slo.88, Luc.93]. The control of flexible manipulators has gained considerable interest over last decade, especially in the space applications. For spacecraft, most of the existing literature addresses the problems related to attitude regulation [Mor.68, Wie.89, Wen.91], stable slewing maneuvers [Wie.85] , and time-optimal or minimum-time slewing maneuvers [Jun.86, Vad.95, Car.93]. Also, this literature considers single-body, and in some cases, rigid spacecraft. However, the tracking control of multibody rigid and flexible spacecraft has not been addressed in the literature in sufficient depth. With the advent of programs such as space station and the possibility of future missions involving space-based multipayload platforms, this has become an imporatant problem. In this chapter, we have attempted to give a partial solution to this very difficult problem.

The main objective of this chapter is to accomplish the asymptotic stability of closed-loop slewing maneuvers of multibody spacecraft while maintaining some minimal tracking performance. The organization of this chapter is as follows. In section 5.2, we address the problem of tracking maneuvers for multibody rigid spacecraft. The case of single-body rigid spacecraft is given in subsection 5.2.1 as a special case of multibody system. Finally, the case of multibody flexible spacecraft is dicussed in section 5.3.

5.2 Tracking of Multibody Rigid Spacecraft

This section addresses the case of multibody rigid spaceraft and shows that the error dynamics for the tracking error performance are globally asymptotically stable. It is assumed, however, that the inertial parameters (masses and inertias) of the system are known and the measurements of the positions as well as the rates are available for each rigid degree of freedom. To avoid dynamic singularities in the attiutude representation, the quaternion formulation is used for the attitude representation of the central body of the spaceraft.

5.2.1 Tracking Error Dynamics

Equations of motion for a rigid multibody spacecraft are given by the following set of vector matrix differential equations, which can be obtained by setting

$K = 0$ and removing q from Eq. (4.1).

$$\begin{aligned} M(p)\ddot{p} &+ C(p,\dot{p})\dot{p} = u \\ \dot{\hat{\alpha}} &= \frac{1}{2}E^T(\hat{\alpha})\omega \end{aligned} \tag{5.1}$$

where $\dot{p} = \{\omega^T, \dot{\theta}^T\}^T$, ω is the 3−vector of angular velocity of central body, $\dot{\theta}$ is an $(n-3)$-vector of velocities of joints between the bodies, $\hat{\alpha} = (\alpha_4, \overline{\alpha}^T)^T$ is the quaternion vector representing the attitude of central body (see Sec. 4.2.2). E is a 3×4 matrix defined as: $E = [-\overline{\alpha},\ \tilde{\overline{\alpha}} + \alpha_4 I]$, $M(p)$ is symmetric and positive definite $n \times n$ inertia matrix, $C(p,\dot{p})$ is the $n \times n$ coefficient matrix of Coriolis and centrifugal forces, and u is an $n \times 1$ vector of applied torques. It can be verified [Hau.89] that $EE^T = I$.

Let the desired trajectory of the system be given by $(\hat{\alpha}_d^T, \theta_d^T, \dot{p}_d^T, \ddot{p}_d^T)$ where $\dot{p}_d = \{\omega_d^T, \dot{\theta}_d^T\}^T$ and $\ddot{p}_d = \{\dot{\omega}_d^T, \ddot{\theta}_d^T\}^T$. The errors in the actual and the desired states are given as follows

$$\begin{aligned} \Delta\alpha_4 &= \alpha_4 - \alpha_{4d} \\ \Delta\hat{\alpha} &= \hat{\alpha} - \hat{\alpha}_d \\ \Delta\theta &= \theta - \theta_d \\ \Delta\dot{p} &= \dot{p} - \dot{p}_d \\ \Delta\ddot{p} &= \ddot{p} - \ddot{p}_d. \end{aligned} \tag{5.2}$$

Define ω_d to satisfy the equation:

$$\dot{\hat{\alpha}}_d = \frac{1}{2}E^T(\hat{\alpha})\omega_d. \tag{5.3}$$

Given $\hat{\alpha}_d(t)$ and $\hat{\alpha}(t)$, $\omega_d(t)$ can be uniquely determined as:

$$\omega_d = 2E(\hat{\alpha})\dot{\hat{\alpha}}_d. \tag{5.4}$$

We assume that $\hat{\alpha}_d(t)$ and $\theta_d(t)$ are given C^2-functions, and completely determine $\dot{p}_d(t)$ and $\ddot{p}_d(t)$. $\hat{\alpha}(t)$ and $\theta(t)$ can be determined from sensor measurements, so that $p(t)$ can be completely determined.

Substituting Eq. (5.2) in Eq. (5.1), we get

$$M(p)(\ddot{p}_d + \Delta\ddot{p}) + C(p,\dot{p})(\dot{p}_d + \Delta\dot{p}) = u_d + \Delta u \tag{5.5}$$

where $u = u_d + \Delta u$. u_d is the feedforward control input which can be determined by the following set of equations.

$$\begin{aligned} M(p)\ddot{p}_d + C(p,\dot{p})\dot{p}_d &= u_d \\ E(\hat{\alpha})\dot{\hat{\alpha}}_d &= \frac{1}{2}\omega_d. \end{aligned} \tag{5.6}$$

It is assumed that $\hat{\alpha}_d$ is chosen to satisfy the quaternion norm constraint

$$\alpha_{4d}^2 + \overline{\alpha}_d^T \overline{\alpha}_d = 1. \tag{5.7}$$

From Eqs. (5.6) and (5.5), the error system equations can be given as follows:

$$\begin{aligned} M(p)\Delta\ddot{p} + C(p,\dot{p})\Delta\dot{p} &= \Delta u \\ \Delta\dot{\hat{\alpha}} &= \frac{1}{2}E^T(\hat{\alpha})\Delta\omega \end{aligned} \tag{5.8}$$

where $\Delta\omega = \omega - \omega_d$.

5.2.2 Stability of Error System

In order to analyze the stability of error system (5.8) with state vector $(\Delta\dot{p}^T, \Delta\theta^T, \Delta\hat{\alpha}^T)^T$, consider a candidate Lyapunov function

$$V = \Delta\dot{p}^T M \Delta\dot{p} + \nu(\Delta\alpha_4)^2 + \Delta\overline{\alpha}^T G_{p1} \Delta\overline{\alpha} + \Delta\theta^T G_{p2} \Delta\theta \tag{5.9}$$

where, ν is a positive scalar and G_{p1} and G_{p2} are 3×3 and $(n-3) \times (n-3)$ symmetric, positive definite matrices, respectively. Note that V is positive definite and radially unbounded in error state vector $(\Delta\hat{\alpha}^T, \Delta\theta^T, \Delta\dot{p}^T)^T$. Taking the time derivative of V yields

$$\dot{V} = 2\Delta\dot{p}^T M \Delta\ddot{p} + \Delta\dot{p}^T \dot{M} \Delta\dot{p} + 2\nu\Delta\alpha_4\Delta\dot{\alpha}_4 + 2\Delta\dot{\overline{\alpha}}^T G_{p1}\Delta\overline{\alpha} + 2\Delta\dot{\theta}^T G_{p2}\Delta\theta. \tag{5.10}$$

Substituting Eq. (5.8) in (5.10)

$$\dot{V} = 2\Delta\dot{p}^T(\Delta u - C\Delta\dot{p} + \frac{1}{2}\dot{M}\Delta\dot{p}) + 2\nu\Delta\alpha_4\Delta\dot{\alpha}_4 + 2\Delta\dot{\overline{\alpha}}^T G_{p1}\Delta\overline{\alpha} + 2\Delta\dot{\theta}^T G_{p2}\Delta\theta. \tag{5.11}$$

Using property $\mathcal{S}$ (Sec. 4.2.1) of the system Eq. (5.11) can be simplified as

$$\dot{V} = 2\Delta\dot{p}^T\Delta u + 2\nu\Delta\alpha_4\Delta\dot{\alpha}_4 + 2\Delta\dot{\overline{\alpha}}^T G_{p1}\Delta\overline{\alpha} + 2\Delta\dot{\theta}^T G_{p2}\Delta\theta. \tag{5.12}$$

Defining $\overline{G}_p = \text{diag}\{\nu, G_{p1}\}$ and rearranging terms, Eq. (5.12) can be rewritten as

$$\dot{V} = \Delta\dot{p}^T(2\Delta u) + 2\Delta\dot{\hat{\alpha}}^T\overline{G}_p\Delta\hat{\alpha} + 2\Delta\dot{\theta}^T G_{p2}\Delta\theta. \tag{5.13}$$

Substituting for $\Delta\dot{\hat{\alpha}}^T$ from Eq. (5.8) in Eq. (5.13), we get

$$\dot{V} = \Delta\dot{p}^T(2\Delta u) + \Delta\omega^T E\overline{G}_p\Delta\hat{\alpha} + 2\Delta\dot{\theta}^T G_{p2}\Delta\theta. \tag{5.14}$$

If Δu is chosen as

$$\Delta u = \begin{bmatrix} -\frac{1}{2}E\overline{G}_p & 0 \\ 0 & -G_{p2} \end{bmatrix} \begin{Bmatrix} \Delta\hat{\alpha} \\ \Delta\theta \end{Bmatrix} - G_r\Delta\dot{p} \tag{5.15}$$

then, substituting for Δu in Eq. (5.14), we get

$$\dot{V} = -2\Delta\dot{p}^T G_r\Delta\dot{p} \leq 0 \tag{5.16}$$

i.e., $\dot{V}$ is negative semidefinite and we can conclude that the error system is at least Lyapunov stable. Also, $\dot{V} = 0 \Rightarrow \Delta\dot{p} = 0$ (since G_r is symmetric and positive definite) $\Rightarrow \omega = \omega_d$, $\dot{\theta} = \dot{\theta}_d$, and $\Delta\ddot{p} = 0$. Substituting $\Delta\dot{p} = \Delta\ddot{p} = 0$ in Eq. (5.8) yields $\Delta u = 0$, i.e., from Eq. (5.15), we get

$$\begin{bmatrix} -\frac{1}{2}E\overline{G}_p & 0 \\ 0 & -G_{p2} \end{bmatrix} \begin{Bmatrix} \Delta\hat{\alpha} \\ \Delta\theta \end{Bmatrix} = 0 \tag{5.17}$$

$$\Rightarrow \Delta\theta = 0 \;\text{ and }\; E\overline{G}_p\Delta\hat{\alpha} = 0. \tag{5.18}$$

Suppose we choose $G_{p1} = \nu I_3$ then $\overline{G}_p = \nu I_4$, and

$$E\overline{G}_p\Delta\hat{\alpha} = \nu E\Delta\hat{\alpha} = 0 \;\Rightarrow\; E\Delta\hat{\alpha} = 0 \tag{5.19}$$

i.e.,

$$[-\overline{\alpha},\ \tilde{\overline{\alpha}} + \alpha_4 I]\begin{Bmatrix} \alpha_4 - \alpha_{4d} \\ \overline{\alpha} - \overline{\alpha}_d \end{Bmatrix} = 0 \tag{5.20}$$

$$\Rightarrow -\overline{\alpha}(\alpha_4 - \alpha_{4d}) + (\tilde{\overline{\alpha}} + \alpha_4 I)(\overline{\alpha} - \overline{\alpha}_d) = 0. \tag{5.21}$$

After some cancellations, we get

$$\begin{aligned} \alpha_{4d}\overline{\alpha} - \tilde{\overline{\alpha}}\,\overline{\alpha}_d - \alpha_4\overline{\alpha}_d &= 0 \\ (\alpha_{4d}\overline{\alpha} - \alpha_4\overline{\alpha}_d) - \tilde{\overline{\alpha}}\,\overline{\alpha}_d &= 0. \end{aligned} \quad (5.22)$$

Let us define vectors v and w as: $v = (\alpha_{4d}\overline{\alpha} - \alpha_4\overline{\alpha}_d)$ and $w = \tilde{\overline{\alpha}}\,\overline{\alpha}_d$. Since vector w is perpendicular to both $\overline{\alpha}$ and $\overline{\alpha}_d$, it is also perpendicular to vector v since v has to be in the plane defined by $\overline{\alpha}$ and $\overline{\alpha}_d$. Then, from Eq. (5.22) and the fact that v and w are orthogonal, it can be concluded that both v and w must be zero vectors, i.e.,

$$v = w = 0. \quad (5.23)$$

In order to determine the conditions under which Eq. (5.23) will be satisfied, we need to examine various possibilities. First, consider the equation $w = 0$. This will be satified under three possible subcases: (a) $\overline{\alpha} = 0$, (b) $\overline{\alpha}_d = 0$, and (c) $\overline{\alpha} \parallel \overline{\alpha}_d$. ($\parallel$ denotes "parallel to".)

In the subcase (a), $\overline{\alpha} = 0$ along with $v = 0 \Rightarrow \alpha_4\overline{\alpha}_d = 0$. But $\overline{\alpha} = 0 \Rightarrow \alpha_4 \neq 0$ (by norm constraint) $\Rightarrow \overline{\alpha}_d = 0 \Rightarrow \Delta\overline{\alpha} = 0 \Rightarrow \Delta\alpha_4 = 0$, and the attitude error is zero along such trajectories. Then by LaSalle's invariance principle error system is asymptotically stable. By similar reasoning as in subcase (a) and interchanging the roles of $\overline{\alpha}$ and $\overline{\alpha}_d$, it can be easily shown that for subcase (b) also $\Delta\hat{\alpha} = 0$, and the error system is again asymptotically stable. In subcase (c), $\overline{\alpha} = \nu\overline{\alpha}_d$, where ν is some scalar. It will be shown later that the only feasible value for this scalar is $\nu = 1$, i.e., $\overline{\alpha} = \overline{\alpha}_d \Rightarrow \alpha_4 = \alpha_{4d} \Rightarrow \Delta\hat{\alpha} = 0$, and the error system is again asymptotically stable.

Now consider the second equation: $v = 0$,

$$v = 0 \Rightarrow \alpha_{4d}\overline{\alpha} - \alpha_4\overline{\alpha}_d = 0 \Rightarrow \overline{\alpha} = (\frac{\alpha_4}{\alpha_{4d}})\overline{\alpha}_d \quad (5.24)$$

provided $\alpha_{4d} \neq 0$. (The case $\alpha_{4d} = 0$ is considered later). If we define $\frac{\alpha_4}{\alpha_{4d}} = k$, Eq. (5.24) can be rewritten as

$$\overline{\alpha} = k\overline{\alpha}_d \quad (5.25)$$

i.e., $\overline{\alpha}$ and $\overline{\alpha}_d$ are collinear. This is same as subcase (c) above where $k = \nu$.

Using Eq.(5.24) in the constraint Eq. 4.9, we get

$$\alpha_4^2 + (\frac{\alpha_4}{\alpha_{4d}})^2 \overline{\alpha}_d^T \overline{\alpha}_d = 1 \tag{5.26}$$

$$\alpha_4^2 + \frac{\alpha_4^2}{\alpha_{4d}^2}(1 - \alpha_{4d}^2) = 1 \tag{5.27}$$

$$\frac{\alpha_4^2}{\alpha_{4d}^2} = 1 \Rightarrow \alpha_4^2 = \alpha_{4d}^2 \tag{5.28}$$

$$\Rightarrow \alpha_4 = \pm\alpha_{4d} \Rightarrow k = \pm 1. \tag{5.29}$$

Substituting Eq. (5.29) in (5.24) yields

$$\overline{\alpha} = \pm\overline{\alpha}_d. \tag{5.30}$$

In Eq. (5.29), if $k = 1 \Rightarrow \overline{\alpha} = \overline{\alpha}_d \Rightarrow \Delta\hat{\alpha} = 0$. For examining the case when $k = -1$, first recall the quaternion equations:

$$\overline{\alpha} = \overline{\alpha}_u \sin(\frac{\phi}{2}); \quad \alpha_4 = \cos(\frac{\phi}{2}) \tag{5.31}$$

$$\overline{\alpha}_d = \overline{\alpha}_{ud} \sin(\frac{\phi_d}{2}); \quad \alpha_{4d} = \cos(\frac{\phi_d}{2}) \tag{5.32}$$

where $\overline{\alpha}_u$ and $\overline{\alpha}_{ud}$ are unit vectors along $\overline{\alpha}$ and $\overline{\alpha}_d$, respectively. For the case $k = -1$, from Eqs. (5.31) and (5.32), we get

$$\alpha_4 = -\alpha_{4d} \Rightarrow \overline{\alpha} = -\overline{\alpha}_d \Rightarrow \frac{\phi}{2} = \pi \pm \frac{\phi_d}{2} \Rightarrow \phi = 2\pi \pm \phi_d \tag{5.33}$$

i.e., ϕ and ϕ_d represent the same orientations in the physical space. Similarly, Eq. (5.33) implies

$$\overline{\alpha} = \sin(\frac{2\pi \pm \phi_d}{2}). \tag{5.34}$$

However, $\overline{\alpha} = -\overline{\alpha}_d$, i.e., in Eq. (5.34) only '+' sign is feasible. This means that ϕ and ϕ_d represent the same orientation in physical space and once again we obtain $\Delta\hat{\alpha} = 0$.

For $v = 0$, there are two more possible subcases: (d) $\alpha_{4d} = 0$ which implies $\alpha_d \neq 0 \Rightarrow \alpha_4 = 0$ (because $v = 0$), and (e) $\overline{\alpha} = \overline{\alpha}_d = 0$. Consider the subcase (d). $\alpha_4 = \alpha_{4d} = 0 \Rightarrow \Delta\alpha_4 = 0$, and from Eq. (5.22)

$$\tilde{\overline{\alpha}}\,\overline{\alpha}_d = 0 \Rightarrow \overline{\alpha} = k\overline{\alpha}_d. \tag{5.35}$$

Again, as shown previously, only feasible value for k is ± 1 which corresponds to the same orientation in the physical space. Similarly for subcase (e) it can be shown, as in subcases (a) and (b), that $\Delta\hat{\alpha} = 0$ and the error system is asymptotically stable.

5.2.3 Special Case: Single-body Rigid Spacecraft

The rotational tracking maneuvers of a single-body rigid spacecraft can be considered as a special case of the multibody spacecraft. The rotational equations of motion for a single-body rigid spacecraft are given by

$$\begin{aligned} J\dot{\omega} + (-\widetilde{J\omega})\omega &= u \\ \dot{\hat{\alpha}} &= \frac{1}{2}E^T\omega \end{aligned} \tag{5.36}$$

where J is moment of inertia matrix of the spacecraft, ω is the angular velocity vector, $\hat{\alpha}$ is the unit quaternion, and u is the vector of applied torques. Note that Eq. (5.36) is written in a form that resembles Eq. (5.1) where matrices $M(p)$ and $C(p,\dot{p})$ are replaced by J and $(-\widetilde{J\omega})$, respectively. Following the same approach as in Sec. 5.2, the error system equations can be given as

$$J\Delta\dot{\omega} + (-\widetilde{J\omega})\Delta\omega = \Delta u \tag{5.37}$$

$$\Delta\dot{\hat{\alpha}} = \frac{1}{2}E^T\Delta\omega \tag{5.38}$$

where $\Delta\omega = \omega - \omega_d$, $\Delta\hat{\alpha} = \hat{\alpha} - \hat{\alpha}_d$, $u = u_d + \Delta u$. It is assumed that the feedforward control u_d can be computed using the following set of equations.

$$u_d = J\dot{\omega}_d + (-\widetilde{J\omega})\omega_d \tag{5.39}$$

$$\omega_d = 2E(\hat{\alpha})\dot{\hat{\alpha}}_d \tag{5.40}$$

where, $\hat{\alpha}_d$ trajectories are designed to satisfy the norm constraint (5.7). Consider a nonlinear control law Δu is given by

$$\Delta u = -\frac{1}{2}E\overline{G}_p\Delta\hat{\alpha} - G_r\Delta\omega. \tag{5.41}$$

As in the multibody case, this control law can be shown to provide global asymptotic stability of the error system (5.37-5.38) if $\overline{G}_p = \nu I_4$ ($\nu > 0$) and G_r is a 3×3 symmetric positive definite matrix.

In [Vad.95], a similar result was published addressing asymptotic stability of tracking error system for a single-body spacecraft, however, the control law given in [Vad.95] was slightly different than Eq. (5.41). The difference in the control laws can be attributed to the difference in the Lyapunov functions used for the stability analysis and the way feedforward control input, u_d, was generated.

5.3 Tracking of Multibody Flexible Spacecraft

The method presented in Sec. 5.2 cannot be applied to the case of *flexible* multibody systems. The reason is that it is not possible to compute $M(p)$ and $C(p, \dot{p})$, and hence u_d, as p includes the flexible mode amplitudes, which cannot be measured or determined. However, if the system is in the attitude-hold configuration, we can, to a limited extent, address the problem of trajectory tracking for multibody flexible spacecraft.

It was shown in Sec. 4.4 that under the static dissipative control law, closed-loop set point maneuvers can be achieved with global asymptotic stability. However, the problem of tracking a given trajectory during such a maneuver was not discussed. Let us suppose that the system is required to follow certain desired trajectory as closely as possible before reaching the final desired orientation. One logical way of modifying the control law given in Eq. (4.34) is to feed back the position and velocity *errors* instead of the position and velocity measurements, and to check if the system will track the desired trajectory with sufficiently small tracking error. In this section, it will be shown that under this modified static disspative control law, system will have some stability properties, although so far it has not been possible to show asymptotic stability of the tracking error. The significance of this result is that the control law is still robust to unmodeled dynamics and parametric uncertainties and it only depends on the measured errors between the actual and the desired trajectories. Also, if the desired trajectories are defined in such a way that the final steady state values are zero for both position- and rate-vectors, the proposed control law will move the system state to the desired state asymptotically. However, no general statements can be made about the tracking performance. If the desired trajectories are such that they do not satisfy the above mentioned condition, then only boundedness of the trajectories is assured. Their asymptotic convergence to the desired state cannot be guaranteed.

Consider a multibody flexible spacecraft in attitude-hold configuration. The equations of motion for such spacecraft (Eq. 4.33) are rewritten as follows.

$$\dot{x} = \begin{bmatrix} 0 & I \\ -M^{-1}K & -M^{-1}(C+D) \end{bmatrix} x + \begin{Bmatrix} 0 \\ M^{-1}B^T \end{Bmatrix} u \tag{5.42}$$

where $x = \{p^T, \dot{p}^T\}^T$ and $p = \{\gamma^T, \theta^T, q^T\}^T$. The closed-loop system with static dissipative control law, Eq. (4.34), can be given as

$$\dot{x} = \begin{bmatrix} 0 & I \\ -M^{-1}\overline{K} & -M^{-1}(C+\overline{D}) \end{bmatrix} x + \begin{Bmatrix} 0 \\ M^{-1}B^T \end{Bmatrix} u \tag{5.43}$$

where, $\overline{K} = (K + B^T\overline{G}_pB)$ and $\overline{D} = (D + B^TG_rB)$.

Now consider a modified static dissipative control law where instead of feeding back the sensed position and rate values, the errors between the actual and the desired values are used for feedback, i.e.,

$$u = -\overline{G}_p(y_p - y_{pd}) - G_r(y_r - y_{rd}). \tag{5.44}$$

where $y_{pd}(t)$, $y_{rd}(t)$ denote the desired position and rate trajectories, respectively. Substituting Eq. (5.44) in Eq. (5.42) yields

$$\dot{x} = \begin{bmatrix} 0 & I \\ -M^{-1}\overline{K} & -M^{-1}(C+\overline{D}) \end{bmatrix} x + \begin{Bmatrix} 0 \\ M^{-1}B^T \end{Bmatrix} d_u \tag{5.45}$$

where d_u is given by

$$d_u = \overline{G}_p y_{pd} + G_r y_{rd}. \tag{5.46}$$

In concise notation, Eq. (5.45) can be rewritten as

$$\dot{x} = f(x) + g(x)d_u. \tag{5.47}$$

Thus, d_u can be considered as an exogenous input to the closed-loop system (5.45). We will show that in the presence of a bounded disturbance d_u, the closed-loop system remains $\mathcal{L}_2$-stable.

For the system (5.47), consider a storage function

$$V(p,\dot{p}) = \frac{1}{2}[p^T, \ \dot{p}^T] \begin{bmatrix} \overline{K} & 0 \\ 0 & M(p) \end{bmatrix} \begin{Bmatrix} p \\ \dot{p} \end{Bmatrix}. \tag{5.48}$$

The time derivative of V is given by

$$\dot{V} = \frac{\partial V}{\partial x}\dot{x} = \nabla V^T \dot{x} \tag{5.49}$$

where ∇V denotes $\frac{\partial V}{\partial x}$. Substituting Eq. (5.47) in Eq. (5.49), we get

$$\dot{V}(x) = \nabla V^T \dot{x} \ = \ \nabla V^T (f(x) + g(x)d_u) \tag{5.50}$$

$$= \ \nabla V^T f(x) + [\nabla V^T g(x)]d_u. \tag{5.51}$$

Noting that $\nabla V^T f(x)$ is simply the time derivative of storage function for system (5.43), we get (with $d_u = 0$)

$$\nabla V^T f(x) \ = \ \dot{V} \tag{5.52}$$

$$= \ -\dot{p}^T \overline{D}\dot{p} \tag{5.53}$$

$$\leq \ -\dot{p}^T B^T G_r B\dot{p} \tag{5.54}$$

$$\leq \ -||y_r||_2^2 \lambda_{min}(G_r) \tag{5.55}$$

where $B\dot{p} = y_r$. Also, ∇V is given by

$$\nabla V = \begin{Bmatrix} \frac{\partial V}{\partial p} \\ \frac{\partial V}{\partial \dot{p}} \end{Bmatrix} = \begin{Bmatrix} \nabla V_1 \\ M(p)\dot{p} \end{Bmatrix} \tag{5.56}$$

where

$$\nabla V_1 = \overline{K}(p)p + \frac{1}{2}\frac{\partial}{\partial p}[\dot{p}^T M(p)\dot{p}]. \tag{5.57}$$

From Eqs. (5.56) and (5.57),

$$\nabla V^T g(x) = [\nabla V_1^T, \ \dot{p}^T M(p)] \begin{bmatrix} 0 \\ M^{-1}(p)B^T \end{bmatrix} = \dot{p}^T B^T = y_r^T. \tag{5.58}$$

Substituting Eqs. (5.58) and (5.55) in Eq. (5.51) yields

$$\dot{V} \leq -||y_r||_2^2 \lambda_{min}(G_r) + y_r^T d_u. \tag{5.59}$$

Integrating Eq. (5.59) from 0 to T gives

$$V(T) - V(0) \leq -||y_r||_T^2 \lambda_{min}(G_r) + < y_r, \ d_u >_T . \tag{5.60}$$

After rearranging the terms

$$0 \leq V(T) \leq V(0) - ||y_r||_T^2 \lambda_{min}(G_r) + < y_r,\ d_u >_T \tag{5.61}$$

$$\Rightarrow ||y_r||_T^2 \lambda_{min}(G_r) \leq \ V(0) + < y_r,\ d_u >_T . \tag{5.62}$$

i.e., the system (5.47) is output strictly passive (OSP). Since OSP property implies finite gain property [Hil.94], it can be concluded that if the exogenous input to the system (d_u) is a square integrable function of t, then the rate output will also be square integrable, i.e., the system output will have finite energy, and stability will be maintained in the $\mathcal{L}_2$- sense. It is to be noted, however, that the asymptotic convergence of position and rate errors to zero cannot be guaranteed. The results obtained in this section are rather limited in scope. Considerable further effort is necessary in order to obtain tracking control laws for multibody flexible systems, which satisfy given performance specifications.

Bibliography

[And.67] Anderson, B. D. O.: A System Theory Criterion for Positive-Real Matrices, *SIAM J. Control*, Vol. 5, 1967.

[Ano.87] General Electric Company Astro-Space Division, "Upper Atmosphere Research Satellite Project Data Book," NASA Goddard Space Flight Center, April 1987.

[Ano.89] "DADS User's Manual", Computer Aided Design Software, Inc., Oakdale, Iowa 52319, July 1989.

[Ano.92] "MSC/Nastran", Software for Finite Element Analysis by MacNeal-Schwendler Corporation, 815 Colorado Boulevard, Los Angeles, CA 90041, 1992.

[Asr.93] Asrar, G. and Dokken, D., editors; EOS Reference Handbook, NP-202, March 1993.

[Bal.82] Balas, M.J.: Trends in Large Space Structures Control Theory: Fondest Hopes, Wildest Dreams, *IEEE Trans. Auto. Contr.*, vol. AC-27, No. 3, 1982.

[Bal.92] Balakrishnan, A.V.: Combined Structures-Controls Optimization of Lattice Trusses, *Computer Methods in Applied Mechanics and Engineering*, vol. 94, 1994.

[Bal.94] Balas, G. J. and Doyle, J. C.: Control of Lightly Damped Flexible Modes in the Controller Crossover Region, *J. Guidance, Control, and Dynamics*, Vol. 17, No. 2, pp. 370-377, 1994.

[Ben.81] Benhabib, R. J., Iwens, R. P., and Jackson, R. L.: Stability of Large Space Structure Control Systems Using Positivity Concepts, *J. Guidance and Control*, Vol. 4, No.5, pp. 487-494, 1981.

[Boo.90] Book, W. J.: Modeling, Design, and Control of Flexible Manipulator Arms: A tutorial Review, Proc. 29th IEEE Conference on Decision and Control, Vol. 2, pp. 500-506, 1990.

[Byr.91] Byrnes, C. I., Isidori, A., and Willems, J. C.: Passivity, Feedback Equivalence, and the Global Stabilization of Minimum Phase Nonlinear Systems, *IEEE Transactions on Automatic Control*, Vol. 36, No. 11, pp. 1228-1240, November 1991.

[Car.93] Carter, M. T., Vadali, S. R., and Singh, T.: Near-Minimum-Time Maneuvers of Large Structures Using Parameter optimization, AIAA Paper 93-3714-CP, pp. 127-137, Guidance, Navigation, and Control Conference, Monterey CA, August 9-11, 1993.

[Des.75] Desoer, C. A., and Vidyasagar, M.: *Feedback Systems: Input-Output Properties*. New York: Academic Press, 1975.

[Doy.82] Doyle, J.C.: Analysis of Feedback Systems With Structured Uncertainty. *IEEE Proceedings*, vol. 129D, No. 6, 1982.

[Fu.87] Fu, K. S., Gonzalez, R. C., and Lee, C. S. G.: *Robotics: Control, Sensing, Vision, and Intelligence*. New York: McGraw Hill, 1987.

[Gre.88] Greenwood, D. T.: *Principles of Dynamics*. New Jersey: Prentice-Hall, 1988.

[Gre.95] Green, M., and Limebeer, D.J.N.: *Linear Robust Control.* New Jersey: Prentice-Hall, 1995.

[Gup.94] Gupta, S.: *State Space Characterization and Robust Stabilization of Dissipative Susyems.* D.Sc. Thesis, George Washington University, 1994.

[Har.64] Harding, C. F.: Solution to Euler's Gyrodynamics, *Journal of Applied Mechanics*, pp. 325-328, June 1964.

[Hau.89] Haug, E. J.: *Computer-Aided Kinematics and Dynamics of Mechanical Systems.* Allyn and Bacon Series in Engineering, 1989.

[Hil.76] Hill, D. and Moylan, P. J.: Stability of Nonlinear Dissipative Systems, *IEEE Transactions on Automatic Control*, Vol. AC-21, No.5, pp.708-711, October 1976.

[Hil.77] Hill, D. J. and Moylan, P. J.: Stability Results for Nonlinear Feedback Systems, *Automatica*, Vol. 13, No. 4, pp. 377-382, 1977.

[Hil.92] Hill, D. J.: Dissipative Nonlinear Systems: Basic Properties and Stability Analysis, Proc. 31st IEEE Conference on Decision and Control, Tucson, Arizona, pp. 3259-3264, December 1992.

[Hil.94] Hill, D., Ortega, R., and van der Schaft A.: *Nonlinear Controller Design Using Passivity and Small-Gain Techniques*, Notes from 1994 IEEE CDC Tutorial Workshop.

[Hyl.93] Hyland, D.C., Junkins, J.L., and Longman, R.W.:. Active Control Technology for Large Space Structures. *J. Guidance, Control, and Dynamics*, Vol. 16, No. 5, 1993.

[Ick.70] Ickes, B. P.: A New Method for Performing Control System Attitude Computation Using Quaternions, *AIAA Journal*, Vol. 8, pp. 13-17, January 1970.

[Ioa.87] P. Ioannou and G. Tao: Frequency-Domain Conditions for Strictly Positive Real Functions, *IEEE Transactions on Automatic Control*, Vol. 32, pp.53-54, Jan. 1987.

[Isi.89] Isidori, A.: *Nonlinear Control Systems*, 2nd ed. Berlin: Springer-Verlag, 1989.

[Jan.93] *Jane's Space Directory*, Jane's Data Division, Jane's Information Group Limited, Sentinel House, 163 Brighton Road, Coulsdon, Surrey CR5 2NH, UK, 1993.

[Jos.89] Joshi, S. M.: *Control of Large Flexible Space Structures.* Berlin: Springer-Verlag, 1989 (Vol. 131, Lecture Notes in Control and Information Sciences).

[Jos.94] Joshi, S. M. and Gupta, S.: Robust Stabilization of Marginally Stable Positive-Real Systems, NASA TM - 109136, July 1994.

[Jos.95] Joshi, S. M., Kelkar, A. G., and Maghami, P. G.: A Class of Stabilizing Controllers for Flexible Multibody Systems, NASA Technical Paper - 3494, May 1995.

[Jos.95a] Joshi, S. M., Kelkar, A. G., and Wen, J. T.-Y.: Robust Attitude Stabilization of Spacecraft Using Nonlinear Quaternion Feedback, *IEEE Transactions on Automatic Control*, Vol. 40, No. 10, pp. 1800-1803, October 1995.

[Jos.95b] Joshi, S. M., Maghami P. G., and Kelkar A. G.: Design of Dynamic Dissipative Compensators for Flexible Space Structures, *IEEE Transactions on Aerospace and Electronic Systems*, Vol. 31, No. 4, pp.1314-1324, October 1995.

[Jos.96] Joshi, S. M. and Gupta, S.: On a Class of Marginally Stable Positive-Real Systems, *IEEE Transactions on Automatic Control*, Vol. 41, No. 1, pp.152-155, January 1996.

[Jua.93] Juang, J.-N., Wu, S.-C., Phan, M., and Longman, R. W.: Passive Dynamic Controllers for Nonlinear Mechanical Systems. *J. Guidance, Control, and Dynamics*, Vol. 16, No.5, pp. 845-851, Sept.-Oct. 1993.

[Jun.86] Junkins, J. L., and Turner, J. D.: *Optimal Spacecraft Rotational Maneuvers*, Amsterdam: Elsevier Scientific, 1986.

[Jun.93] Junkins, J. L. and Kim, Y.: *An Introduction to Dynamics and Control of Flexible Structures*, Washington, DC: AIAA, 1993.

[Kai.80] T. Kailath: *Linear Systems.* N.J: Prentice-Hall, 1980.

[Kan.73] Kane, T. R.: Solution of Kinematical Differential Equations for a Rigid Body, *Journal of Applied Mechanics*, pp. 109-113, March 1973.

[Kau.94] H. Kaufman, I. Bar-Kana, and K. Sobel: *Direct Adaptive Control Algorithms: Theory and Applications.* New York: Springer Verlag, 1994.

[Kel.94] Kelkar A. G.: Mathematical Modeling of a Class of Multibody Flexible Space Structures", NASA Technical Memorandum - 109166, December 1994.

[Kel.95] Kelkar, A. G., Joshi, S. M., and Alberts, T. E.: Dissipative Controllers for Nonlinear Multibody Flexible Space Systems, *Journal of Guidance, Control, and Dynamics*, Vol. 18, N0. 5, pp. 1044-1052, 1995.

[Kel.95a] Kelkar, A. G. and Joshi, S. M.: Global Stabilization of Multibody Spacecraft Using Quaternion-Based Nonlinear Control Law, Proc. 1995 American Control Conference, The Westin Hotel, Seattle, Washington, June 21-23, Vol. 5, pp. 3612-3615, 1995.

[Kod.84] Koditschek, D. E.: Natural Control of Robot Arms, Proc., IEEE Conference on Decision and Control, Las Vegas, Nevada., pp. 733-735, 1984.

[Lim.92] Lim, K.B., Maghami, P.G., and Joshi, S.M.: Comparison of Controller Designs for an Experimental Flexible Structure. *IEEE Control Systems Magzine*, Vol. 12, No. 3, pp. 108-118, 1992.

[Loz.90] Lozano-Leal, R., and Joshi, S. M.: Strictly Positive Real Functions Revisited. *IEEE Trans. Auto. Control*, Vol. 35, No. 11, pp.1243-1245, November 1990.

[Luc.93] Lucibello, P. and Di Benedetto,M. D.: Output Tracking for a Nonlinear Flexible arm, *Journal of Dynamic Systems, Measurement and Control*, Vol. 115, pp. 78-85, March 1993.

[McL.87] McLaren, M. D., and Slater, G. L.: Robust Multivariables Control of Large Space Structures Using Positivity. *J.Guidance, Control and Dynamics*, Vol. 10, pp. 393-400, July-Aug. 1987.

[Mei.70] Meirovitch, L.: *Methods of Analytical Dynamics.* New York: McGraw Hill, 1970.

[Mei.90] Meirovitch, L.: *Dynamics and Control of Structures.* New York: John Wiley, 1990.

[Mor.68] Mortesen, R. E.: A Globally Stable Linear Attitude Regulator, *International Journal of Control*, Vol. 8, No. 3, pp. 297-302, 1968.

[New.66] Newcomb, R. W.: *Linear Multiport Synthesis*, New York: McGraw-Hill, 1966.

[Ng.88] Ng, A. C. and Modi, V. J.: A Formulation for Studying Dynamics of Interconnected Bodies With Application, Proc. AIAA/AAS Astrodynamics Conference, Paper No. 88-4303-CP, pp. 660-668, August 15-17, 1988.

[Nij.90] Nijmeijer, H. and van der Schaft, A. J.: *Nonlinear Dynamical Control Systems.* New York: Springer Verlag, 1990.

[Nur.84] Nurre, G.S., Ryan, R.S., Scofield, H.N., and Sims, J.L.: Dynamics and Control of Large Space Structures. *J. Guidance, Control, and Dynamics*, vol. 7, No. 5, 1984.

[Pac.93] Packard, A.K., Doyle, J.C., and Balas, G.J.: Linear Multivariable Robust Control With a μ-Perspective. *ASME J. Dynamic Systems, Measurement, and Control*, vol. 115 No. 2(B), 426-438, 1993.

[Pad.88] Paden, B., and Panja, R.: Globally Asymptotically Stable PD+ Controller for Robot Manipulators. *Int. Journal of Control*, Vol. 47, No. 6, pp 1697-1712, 1988.

[Pad.90] Paden, B., Riedle, B., and Bayo, E.: Exponentially Stable Tracking Control for Multi-Joint Flexible-Link Manipulators. Proc. 1990 American Control Conference, San Diego, California, May 23-25, pp. 680-684, 1990.

[Pop.73] Popov, V. M.: *Hyperstability of Control Systems.* Berlin: Springer-Velag, pp. 118-230, 1973.

[Rus.80] Russell, R.A., Campbell, T.G., and Freeland, R.E.: A Technology Development Program for Large Space Antenna. NASA TM-81902, 1980.

[Sin.85] Singh, S. N. and Schy, A. AA.: Robust Trajectory Following Control of Robotic Systems, *Journal of Dynamic Systems, Measurement and Control*, Vol. 107, pp.308-314, December 1985.

[Sin.89] Singh, G., Kabamba, P. T., and McClamroch, N. H.: Planar, Time-Optimal, Rest-to-Rest Slewing Maneuvers of Flexible Spacecraft, *J. Guidance, Control, and Dynamics*, Vol. 12, No. 1, pp.71-81, 1989.

[Sin.90] Singh, G., Kabamba, P. T., and McClamroch, N. H.: Bang-Bang Control of Flexible Spacecraft Slewing Maneuvers: Guaranteed Terminal Point Accuracy, *J. Guidance, Control, and Dynamics*, Vol. 13, No. 2, pp.376-379, 1990.

[Sla.90] Slater, G. L., and McLaren, M. D.: Estimator Eigenvalue Placement in Positive Real Control. *J. Guidance, Control and Dynamics*, Vol. 13, No.1, pp.168-175, 1990.

[Slo.88] Slotine, J.J. E. and Li, W.: Adaptive Manipulator Control: A Case Study, *IEEE Transactions on Automatic Control*, Vol.33, No.11, pp.995-1003, November 1988.

[Spo.89] Spong, M. W. and Vidyasagar, M.: *Robot Dynamics and Control*, New York: John Wiley and Sons, 1989.

[Ste.87] Stein, G., and Athans, M.: The LQG/LTR Procedure for Multivariable Feedback Control Design. *IEEE Trans. Auto. Control*, Vol. 32, No. 2, pp. 105-114, 1987.

[Tak.81] Takegaki, M. and Arimoto, S.: A New Feedback Method for Dynamic Control of Manipulators. *ASME Journal of Dynamic Systems, Measurement and Control*, Vol. 102, June 1981.

[Tao.88] Tao, G. and Ioannou, P.: Strictly Positive Real Matrices and the Lefschetz-Kalman-Yakubovich Lemma, *IEEE Trans. Automat. Contr.*, vol. 33, pp.1183-1185, Dec. 1988.

[Tay.74] Taylor, J. H.: Strictly Positive Real Functions and Lefschetz-Kalman-Yakubovich (LKY) Lemma, *IEEE Trans. Circuits Syst.*, pp. 310-311, March 1974.

[Vad.95] Vadali, S. R., Carter, M. T., Singh, T. and Abhyankar, N. S.: Near-Minimum-Time Maneuvers of Large Structures: Theory and Experiments, *Journal of Guidance, Control and Dynamics*, Vol. 18, No. 6, pp. 1380-1385, 1995

[Van.65] Van Valkenberg, M. E.: *Introduction to Modern Network Synthesis*. New York: John Wiley, 1965.

[Vid.81] Vidyasagar, M.: *Input-Output Analysis of Large-Scale Interconnected Systems*. Berlin: Springer Verlag, 1981. (Vol. 29, Lecture Notes in Control and Information Sciences).

[Vid.93] Vidyasagar, M.: *Nonlinear Systems Analysis*, 2nd ed., Englewood Cliffs, New Jersy: Prentice Hall, 1993.

[Wen.88] Wen, J. T., and Bayard, D. S.: A New Class of Control Laws for Robotic Manipulator. *Int. Journal of Control*, Vol. 47, No. 5, pp. 1361-1385, 1988.

[Wen.88a] Wen, J. T.: Time Domain and Frequency Domain Conditions for Strict Positive Realness, *IEEE Trans. Automat. Contr.*, vol. 33, pp.988-992, Oct. 1988.

[Wen.91] Wen, J. T. and Kreutz-Delgado, K.: The Attitude Control Problem, *IEEE Transactions on Automatic Control*, Vol. 36, No. 10, pp. 1148-1163, 1991.

[Wie.85] Wie, B. and Barba, P. M.: Quaternion Feedback for Spacecraft Large Angle Maneuvers, *Journal of Guidance, Control and Dynamics*, Vol.8, No.3, pp. 360-365, 1985.

[Wie.89] Wie, B., Wiess, H., and Arapostathis, A.: Quaternion Feedback Regulator for Spacecraft Eigenaxis Rotations, *Journal of Guidance, Control and Dynamics*, Vol.12, No.3, pp. 375-380, 1989.

[Wil.72] Willems, J. C.: Dissipative Dynamical Systems-Parts I and II: General Theory, *Arch. Rational Mechanics and Analysis*, Vol. 45, pp. 321-351 and pp. 352-391, 1972.

[Yua.93] Yuan, B. S., Book, W. J., and Huggins, J. D.: Dynamics of Flexible Manipulator Arms: Alternate Derivation, Verification, and Characteristics for Control, *Journal of Dynamic Systems, Measurement, and Control*, Vol. 115, No. 3, pp.394-404, 1993.

Index

Lecture Notes in Control and Information Sciences

Edited by M. Thoma

1993–1996 Published Titles:

Vol. 186: Sreenath, N.
Systems Representation of Global Climate Change Models. Foundation for a Systems Science Approach.
288 pp. 1993 [3-540-19824-5]

Vol. 187: Morecki, A.; Bianchi, G.; Jaworeck, K. (Eds)
RoManSy 9: Proceedings of the Ninth CISM-IFToMM Symposium on Theory and Practice of Robots and Manipulators.
476 pp. 1993 [3-540-19834-2]

Vol. 188: Naidu, D. Subbaram
Aeroassisted Orbital Transfer: Guidance and Control Strategies
192 pp. 1993 [3-540-19819-9]

Vol. 189: Ilchmann, A.
Non-Identifier-Based High-Gain Adaptive Control
220 pp. 1993 [3-540-19845-8]

Vol. 190: Chatila, R.; Hirzinger, G. (Eds)
Experimental Robotics II: The 2nd International Symposium, Toulouse, France, June 25-27 1991
580 pp. 1993 [3-540-19851-2]

Vol. 191: Blondel, V.
Simultaneous Stabilization of Linear Systems
212 pp. 1993 [3-540-19862-8]

Vol. 192: Smith, R.S.; Dahleh, M. (Eds)
The Modeling of Uncertainty in Control Systems
412 pp. 1993 [3-540-19870-9]

Vol. 193: Zinober, A.S.I. (Ed.)
Variable Structure and Lyapunov Control
428 pp. 1993 [3-540-19869-5]

Vol. 194: Cao, Xi-Ren
Realization Probabilities: The Dynamics of Queuing Systems
336 pp. 1993 [3-540-19872-5]

Vol. 195: Liu, D.; Michel, A.N.
Dynamical Systems with Saturation Nonlinearities: Analysis and Design
212 pp. 1994 [3-540-19888-1]

Vol. 196: Battilotti, S.
Noninteracting Control with Stability for Nonlinear Systems
196 pp. 1994 [3-540-19891-1]

Vol. 197: Henry, J.; Yvon, J.P. (Eds)
System Modelling and Optimization
975 pp approx. 1994 [3-540-19893-8]

Vol. 198: Winter, H.; Nüßer, H.-G. (Eds)
Advanced Technologies for Air Traffic Flow Management
225 pp approx. 1994 [3-540-19895-4]

Vol. 199: Cohen, G.; Quadrat, J.-P. (Eds)
11th International Conference on Analysis and Optimization of Systems – Discrete Event Systems: Sophia-Antipolis, June 15–16–17, 1994
648 pp. 1994 [3-540-19896-2]

Vol. 200: Yoshikawa, T.; Miyazaki, F. (Eds)
Experimental Robotics III: The 3rd International Symposium, Kyoto, Japan, October 28-30, 1993
624 pp. 1994 [3-540-19905-5]

Vol. 201: Kogan, J.
Robust Stability and Convexity
192 pp. 1994 [3-540-19919-5]

Vol. 202: Francis, B.A.; Tannenbaum, A.R. (Eds)
Feedback Control, Nonlinear Systems, and Complexity
288 pp. 1995 [3-540-19943-8]

Vol. 203: Popkov, Y.S.
Macrosystems Theory and its Applications: Equilibrium Models
344 pp. 1995 [3-540-19955-1]

Vol. 204: Takahashi, S.; Takahara, Y.
Logical Approach to Systems Theory
192 pp. 1995 [3-540-19956-X]

Vol. 205: Kotta, U.
Inversion Method in the Discrete-time Nonlinear Control Systems Synthesis Problems
168 pp. 1995 [3-540-19966-7]

Vol. 206: Aganovic, Z.;.Gajic, Z.
Linear Optimal Control of Bilinear Systems with Applications to Singular Perturbations and Weak Coupling
133 pp. 1995 [3-540-19976-4]

Vol. 207: Gabasov, R.; Kirillova, F.M.; Prischepova, S.V.
Optimal Feedback Control
224 pp. 1995 [3-540-19991-8]

Vol. 208: Khalil, H.K.; Chow, J.H.; Ioannou, P.A. (Eds)
Proceedings of Workshop on Advances in Control and its Applications
300 pp. 1995 [3-540-19993-4]

Vol. 209: Foias, C.; Özbay, H.; Tannenbaum, A.
Robust Control of Infinite Dimensional Systems: Frequency Domain Methods
230 pp. 1995 [3-540-19994-2]

Vol. 210: De Wilde, P.
Neural Network Models: An Analysis
164 pp. 1996 [3-540-19995-0]

Vol. 211: Gawronski, W.
Balanced Control of Flexible Structures
280 pp. 1996 [3-540-76017-2]

Vol. 212: Sanchez, A.
Formal Specification and Synthesis of Procedural Controllers for Process Systems
248 pp. 1996 [3-540-76021-0]

Vol. 213: Patra, A.; Rao, G.P.
General Hybrid Orthogonal Functions and their Applications in Systems and Control
144 pp. 1996 [3-540-76039-3]

Vol. 214: Yin, G.; Zhang, Q. (Eds)
Recent Advances in Control and Optimization of Manufacturing Systems
240 pp. 1996 [3-540-76055-5]

Vol. 215: Bonivento, C.; Marro, G.; Zanasi, R. (Eds)
Colloquium on Automatic Control
240 pp. 1996 [3-540-76060-1]

Vol. 216: Kulhavý, R.
Recursive Nonlinear Estimation: A Geometric Approach
244 pp. 1996 [3-540-76063-6]

Vol. 217: Garofalo, F.; Glielmo, L. (Eds)
Robust Control via Variable Structure and Lyapunov Techniques
336 pp. 1996 [3-540-76067-9]

Vol. 218: van der Schaft, A.
L_2 Gain and Passivity Techniques in Nonlinear Control
176 pp. 1996 [3-540-76074-1]

Vol. 219: Berger, M.-O.; Deriche, R.; Herlin, I.; Jaffré, J.; Morel, J.-M. (Eds)
ICAOS '96: 12th International Conference on Analysis and Optimization of Systems - Images, Wavelets and PDEs: Paris, June 26-28 1996
378 pp. 1996 [3-540-76076-8]

Vol. 220: Brogliato, B.
Nonsmooth Impact Mechanics: Models, Dynamics and Control
420 pp. 1996 [3-540-76079-2]